AA

Glovebox Guide

WILDLIFE OF BRITAIN

Roy Brown

D1512846

Produced by the Publishing Division
of The Automobile Association

4

Editor: *Roger Thomas*
Art Editor: *Harry Williams FCSD*
Illustrations: *Andrew Hutchinson*
Cover picture: *Hazel or Common Dormouse*
Typesetting: *Afal, Cardiff*
Printing: *Purnell Book Production Ltd, a member of the*
BPCC Group

Produced by the Publishing Division of
The Automobile Association

Distributed in the United Kingdom by the
Publishing Division of The Automobile
Association, Fanum House, Basingstoke,
Hampshire RG21 2EA

ISBN 0 86145 686 6

Published by The Automobile Association

Glovebox Guide

WILDLIFE OF BRITAIN

This book describes nearly 70 species of animal living wild in Britain today. It tells you what to look for and where to find animals, with details of their homes, family life, what they eat and the noises they make. It also describes tracks, trails and other clues, which are often easier to find than the animals themselves.

The book is divided into mammals, amphibians and reptiles. Within each category, the animals appear in alphabetical order.

About the author

Roy Brown developed an early interest in natural history as a boy in the East End of London. He studied at Leeds and London Universities, was a University Lecturer and is now Head of Advisory Services at the North York Moors National Park, where he lives with his family. His first book on mammal tracks was published when he was 20 and remained a standard for 20 years; later work includes a study of bird tracks and signs. He is a Fellow of the Institute of Biology, a visiting lecturer at four universities and frequently broadcasts on television and radio.

ACKNOWLEDGEMENTS

The publishers gratefully acknowledge the following for the use of their photographs. Where more than one photographer contributed to a page, credits are listed in order of position of the photographs, top to bottom.
10 Chris Pellant. **11** Roy Brown. **14** Tony Hopkins. **15** Roy Brown, Harry Williams. **18** Roy Brown. **19** Roy Brown, W R Mitchell. **21, 24, 28, 29** Roy Brown. **32** Michael Clark. **33** W R Mitchell. **38** Roy Brown. **39** W R Mitchell. **40** Joyce Pope. **41-4** Roy Brown. **47** W R Mitchell. **55, 56** Nature Photographers Ltd. **61-3** Biophoto Associates. **64, 65, 73, 76, 79, 83** Roy Brown. **97** Chris Pellant. **101** Roy Brown (lizard skin and trail), Biophoto Associates. **102, 103** Biophoto Associates. **104** Nature Photographers Ltd, Biophoto Associates. **105** Biophoto Associates. **108** Roy Brown. **117** S & O Mathews. **118** Tony Hopkins.

WILDLIFE OF BRITAIN

Contents

INTRODUCTION

Since the last glaciers left the land and changes in sea level cut off the British Isles from mainland Europe, the variety of our mammal, reptile and amphibian life has been limited. The climate means that all snakes, lizards, frogs, toads, newts and a few mammals such as the bats, hedgehog and dormouse have to hibernate to survive. Over the centuries people have destroyed some species such as wolves and bears, which were a direct threat, by hunting them down. The range of others like pine marten and wild cat in England has been greatly reduced by the destruction of their habitat. By the same token, introductions, both deliberate and accidental, have enriched our animal life, and human changes to land use, such as the planting of large coniferous forests, the construction of miles of motorway embankment and even the building of cities have created favourable new habitats for some animals. We all take rats, house mice, grey squirrels, rabbits and fallow deer for granted but they have all been introduced in historic times. Moles, once confined to woodland soils, have thrived in the well-drained, worm-rich pastures created by farmers; the pipistrelle bat is almost completely dependent on buildings for its roosts and the polecat, once rare and confined to central Wales, has started to copy the now very common fox in supplementing its diet from dustbins in some areas. The development of our animal populations is very closely linked to humans. Although not very rich when compared with some parts of the world, our animal life is all around us and the careful observer, using this guide, will be able to find out a great deal about animals and their behaviour.

This book deals with three groups of animals, the mammals, the reptiles (snakes and lizards) and amphibians (frogs, toads and newts). Most people have seen some of the more common mammals, such as the rabbit, hedgehog or fox and will be aware of others through their signs, eg mole hills on the lawn. Similarly, almost everyone will be aware of frogs, their spawn and tadpoles, although even these are becoming less common.

Top: *Adder*
Right: *Common toad*
Below: *Pine marten*

However, many <u>species</u> are not very big, have a limited distribution, are secretive or are active only at night and so are rarely seen. Nevertheless, even the smallest creatures leave signs of their activity going about their daily routine of keeping clean, searching for food, breeding or, in the case of some mammals, just playing.

Certain words in the text are underlined (eg <u>species</u> in the above paragraph). Such words are fully explained in the glossary at the back of the book.

ANIMAL STRUCTURES

All the animals described here have a backbone and four limbs (except the snakes). Mammals are warm blooded, covered in fur and give birth to live young. Both the reptiles and amphibians are less able to control their body temperatures and become inactive in cold weather. The reptiles lay eggs, although in some species these hatch in the body and the young appear to be born live. The amphibians spend much of the year on dry land, but have to go back to water to breed; the water is needed to protect the spawn and to allow the young to survive before they <u>metamorphose</u> into the final adult form.

LIMBS

The vast majority of the animals described here walk, run or jump on four feet. The only exceptions are the bats, the snakes and the slow worm (a legless lizard) where there are no external signs of limbs. The variation in the limbs means that each species leaves distinctive <u>tracks</u> in the ground, if it is soft enough. The snakes leave sinuous furrows in soft ground, while lizards and newts, with their arms and legs to the sides of the body, leave indistinct footprints with very clear body and tail marks. The frogs and toads, with their stronger limbs and inwardly turning hands, leave distinct outlines in soft silt.

Mammal limbs, with a basic pattern of five toes on each, show considerable variety. Some species walk on the flat of their feet, like the hedgehog, and are said to be <u>plantigrade.</u> Others, like the fox, walk on their toes and are <u>digitigrade,</u> while some, like the deer, walk on the extreme tips of two toes only and are <u>unguligrade.</u> This progression reflects increasing adaptation for speed. The plantigrade and digitigrade animals leave tracks with pad marks, while the unguligrades leave tracks with <u>slots</u> or <u>cleave</u> marks.

MAMMAL LIMB AND TRACK TYPES

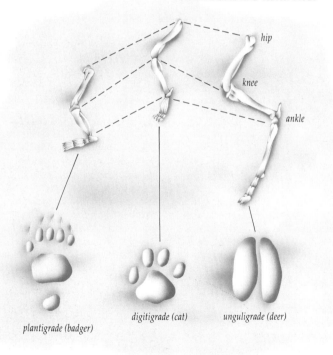

hip

knee

ankle

plantigrade (badger)

digitigrade (cat)

unguligrade (deer)

SKULLS

The skulls of animals, especially the variations in the teeth, tell a great deal about the way of life. The teeth type and pattern varies according to the diet and will tell if the skull belonged to a carnivore, insectivore, herbivore or omnivore.

Skulls of larger mammals may be found years after death and smaller ones often turn up in owl pellets or even droppings. The diagrams below show the main skull types and how to identify them.

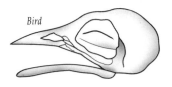
Bird

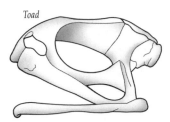
Toad

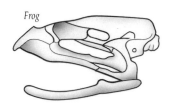
Frog

Mammal

SKULL TYPES

Bird	Teeth absent; beak present.
Toad	Teeth absent; simple unpointed tips on jaws.
Frog	Teeth present; all similar in shape to each other; very small; in upper jaw only.
Lizard	Teeth present; similar in shape to each other; conical.
Snake	Teeth present; similar in shape to each other; sharp and backward pointing.
Mammal	Teeth present; different in shape to each other.

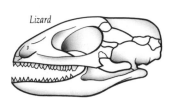
Lizard

Snake

Detailed keys to the different types of mammal skull are contained in some of the books in the bibliography.

ANIMAL ACTIVITY SIGNS

Many animals have a daily routine which leaves a distinctive and instructive series of signs. Wherever you look, you will find evidence of activity, be it mice runs in multi-storey car parks or otter slides on remote river banks in Scotland. Mammals in particular leave a range of signs, including tracks, pathways, feeding signs, grooming and territorial marking signs, homes and droppings. Reptiles leave sloughed skins and, occasionally, tracks on soft sand or damp soil. Amphibians leave few signs, but are more frequently seen in the right

The adder sheds its skin

habitats. The key at all times is to look carefully, not just for the animals themselves but for the many clues to their activity.

TRACKS *(see pgs 112-16)*

When an animal moves across soft ground, such as mud, silt, wet sand, snow or even deep dust it leaves tracks. Depending on the speed, the age and the condition of the individual it will leave a series of tracks and <u>accessory marks</u> from other parts of the body in a <u>trail.</u> When an animal walks, the front and the back limbs on opposite sides are moved at the same time. The <u>stride</u> will be small and the hind foot will be placed almost exactly over the preceding front foot to give <u>registered</u> tracks. As the pace increases to a run the registration becomes less exact and the stride increases. When an animal bounds or gallops all four tracks are distinct, sometimes bunched together in fours with a long stride between each group. Page 17 shows how to record information on tracks for maximum interest.

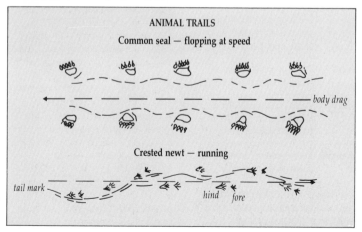

ANIMAL TRAILS

Common seal — flopping at speed

body drag

Crested newt — running

tail mark *hind* *fore*

PATHWAYS AND RUNS

Pathways occur on the ground, in the litter layers and below ground. In woodland undergrowth well-worn paths are formed. These may last for many years and their size is determined by that of the largest user. If a path is used by badgers it will be about 25cm wide and clear of vegetation about 40cm above ground.

A fox will use the same path, but it will be clear to at least 50cm high. Deer paths are narrow; those used by muntjac are clear of vegetation to about 60cm, those used by roe deer to 100cm. In grassland, heavily trodden badger paths are obvious, as are extensive rabbit runs which are no more than 15cm wide. In waterside vegetation chinese water deer make open pathways in the reedbeds, while

Rabbit path through the woods

ducks create smaller tunnels. The water vole has regular runways between the water and its burrow entrances. Some mammals make runs and tunnels below ground. Voles create temporary runs between the snow and the soil, while the mole makes a permanent tunnel system in the soil. In the litter layers of old coarse grass, shrews and voles make extensive tunnel systems, those of the voles containing masses of green, cylindrical droppings and chopped vegetation. The wood mouse burrows in woodland litter. Holes are made in hedge bottoms by rabbit (15—20cm diameter), fox and badger (up to 50cm diameter). In buildings many animals follow obvious pathways. Brown rats create extensive tunnels in old-style baled hay stores. House mice and rats will follow regular routes along rafters or in roof spaces, these being marked by greasy body deposits.

FEEDING REMAINS *(see pgs 81-3)*

Plant Many creatures feed on plant material and leave obvious signs of their activity. Tree roots are gnawed, the bark and even the wood taken by squirrels, rabbits, hares and deer. The green shoots of trees, bushes and undergrowth are browsed by many deer, and the base of the <u>canopy</u> may be cut into a straight line at different heights by deer, goats, rabbits, horses and cattle. Many small mammals and birds are dependent on cones from pine, spruce and fir trees and the nuts and seeds of other trees and plants in the autumn. These are often found in accumulations at favourite feeding places. Each species has a distinctive way of opening these. Soft fruit and berries are also important food sources in the autumn. Fungi are nibbled by rabbits and rodents, which leave distinctive teeth marks. Finally, a number of species graze off ground vegetation, the closely cropped 'lawn' of grass around a water vole hole being a prime example.

Animal The remains of whole carcasses with distinctive teeth marks are left by carnivores like fox, wild cat, stoat and some birds of prey. Crushed bones, the remains of egg shells, and mollusc and crustacean shells which have been crushed or nibbled can all tell you which animal has been there. Feathers are a useful clue. Naturally moulted feathers have intact stem, those plucked by a bird of prey have a split stem, those pulled from a dead bird by a small carnivore, eg a stoat, will have the stems bitten through, while those torn out by a fox will be broken and have damaged plumes which are matted together.

Rabbits are a favourite prey of foxes

DROPPINGS AND MARKING POINTS (see pg 84)

All animals produce waste products. Mammal droppings reflect feeding preferences and many species can be identified by the droppings' shape, content and size. Carnivore droppings often contain bone fragments, fur and feathers. They may be similar to owl

Fox dropping

pellets, but are more twisted and the contents more fragmented with a strong musty smell. Insectivore droppings, hedgehog and bat for example, consist entirely of insect remains. Herbivore droppings, eg rabbit and deer, contain fibrous remains of plants. Reptiles and amphibians have a <u>clocoa</u> where the alimentary tract, urine-carrying and genital ducts open. The droppings formed here are mixed with concentrated urine to give white-capped amorphous masses, which are of little use for species identification. Body smells are another useful sign. The fox produces a pungent, lingering scent from glands over which it has no control. An occupied badger set has a sweet, musty smell, while a deer resting place has a cow-like odour. Many mammals use regular latrines and also deposit urine or scent from body glands to mark their territory. Roe deer score trees with their antlers and scent them with face glands.

PELLETS

These are the undigested remains of prey regurgitated by birds such as owls. They may be found in large numbers below a roost and their contents of bones and skulls will give a great deal of information about the mammal species living in the area.

HOMES

Burrow entrances in banks are the most obvious, but small ball nests in low vegetation (eg harvest mouse), large globular stick nests (dreys) of squirrels in trees and shallow scrapes in ground vegetation (hares) or flattened lying-up places (deer) are obvious if you know where and how to look.

GROOMING AND MISCELLANEOUS SIGNS

Many signs do not fit neatly into one category. Moulted hair (especially tufts on fences), shed antlers, frayed trees where deer rub the dead velvet from their antlers or mark territory, scenting points on boulders or trees, tooth and claw sharpening and cleaning marks on wood, rubbing and scratching posts are all common signs. Digging activity is frequent. Signs of play, such as rolled vegetation or river bank slides, deer <u>rutting</u> scrapes and, occasionally, sloughed skins from lizards and snakes all illustrate the range of material you can come across on a patient search.

HABITATS

Some animals, especially the reptiles and amphibians, are restricted to certain habitats because of particular breeding or feeding requirements, but many species of mammal will occur in a range of environments. The following information is therefore intended only as a guide.

DECIDUOUS WOODLAND

There are many different types of woodland ranging from oak, ash and birch to wet areas dominated by alder and willows. Typical animals include badger, fox, roe and fallow deer, many bats, stoat, mole and hedgehog — although now more commonly associated with farmland, these two latter species were originally creatures of the forest floor. The wood mouse and hazel doormouse are long-established residents, while the grey squirrel is a more recent addition. Common shrews are widespread, especially on the edges of woodland. Grass snakes are also associated with woodland edges.

CONIFEROUS FORESTS

The only native coniferous woodlands are some of the pine forests in Scotland, but planting this century has benefited several species. The red squirrel and pine marten are both tree-top dwellers in the pine. Roe deer have spread greatly in many areas in northern England because of Forestry Commission activity and fallow deer have thrived within this woodland.

MOOR AND HEATH

The high moors of England and Scotland have wild cat and mountain hare. Feral goats, moor sheep and, in some places, ponies are large and obvious mammals, while examination of the ground will reveal common and pygmy shrew. Heath areas are favoured by adders (also found on improved pastures), common and sand lizards. In Scotland both otters and pine martens may be associated with moorland streams. On the wet boggy areas of central Wales the polecat is starting to make a comeback. Red deer are widespread on Scottish moors.

Deciduous woodland at Monk's Wood, Huntingdon

FRESHWATER AREAS

Ponds are the breeding places for frogs, toads and newts. Well-covered river banks, especially in wooded areas, may be used by the otter (now very rare in England and Wales), mink and water vole. The distinctive black and white water shrew may be seen on still and moving water, but in common with many others it is more likely to be identified from the signs it leaves than from a sighting.

PASTURES

These are traditional grasslands, ranging from limestone turfs where slow worm and adder are found to coarse acid grasslands with extensive

The South Downs, West Sussex

field vole and common shrew runs in the rotting vegetation. Moles are common in pasture soils. Bats often hunt over old pastures for the large insects found there, and damp grasslands are very important for the survival of frogs and newts.

COAST

Other than the occasional dolphin or porpoise washed ashore the main mammals to look out for are the common and grey (or Atlantic) seal. Land cliffs and old quarry faces with caves may be important for bats.

FARMLAND

There are many habitats here. In the hedgerows and unploughed field edges, wood mice, house mice, harvest mice, shrews, hedgehogs, stoats, weasels and rabbits are all to be found.

Farm dwellings attract mice, rats, birds and bats

Around farm buildings brown rats and house mice may be common, and roofs may be used as roosts by several species of bat. Owls and kestrels often roost in the outbuildings.

VERGES, EMBANKMENTS AND DISTURBED LAND

These coarse, overgrown areas provide a range of habitats and are a stronghold for badger, harvest mouse, rabbit and many of the small mammals. In their turn they support the larger carnivores. The adder and grass snake are common, along with the kestrel and sparrowhawk.

BUILT-UP AREAS

The urban fox is now a well-known sight. Bats, particularly pipistrelles, nest in house roofs and are often seen hunting around street lamps at night. Although not welcome guests, both house mice and brown rats are still common in some areas. Even multi-storey car parks are frequented by some rodents, bats and birds of prey and are well worth keeping an eye on.

WHERE AND HOW TO LOOK

It is worth looking for signs anywhere where there is cover and/or wet ground. Muddy paths, river banks, mud flats and sand dunes are good. The edges of freshly worked fields and even dust by the side of the road are good for tracks.

Fresh snow is another good surface, but if it is too deep then the tracks will be distorted. Animal paths and runs often lead to <u>latrines</u>, resting

A pine marten's tracks in the snow

areas or burrows. Look in hedge bottoms for runs. Look in the trees for bat roosts in trunk holes, squirrel dreys (nests) in the branches, scratch marks on the trunk and feeding or territorial marking on the trunk and in foliage. River banks and hedgerows are rich in signs, as are lay-bys backing on to woodland or agricultural land.

While tracks and signs may be seen anywhere and at any time of the day it is more difficult, other than by chance sighting, to see the animals themselves. Dusk is perhaps the best time to go looking.

Larger animals may be stalked. Open rides in forestry plantations are good places to look for deer, for example. To stalk an animal follow a number of simple rules:

Move quietly and slowly, wear clothes which blend in with the surroundings and do not rustle.

Do not become silhouetted against the skyline and try to stay downwind from the animal or home you are watching, with the breeze blowing from it to you. Follow the animal's pattern of behaviour. Move when it is feeding, stop when it looks up. Move slowly and smoothly, directly towards the animal. If you are seen, move gently sideways and do not make eye contact. If you are watching a nest or burrow try to get above ground level and make yourself comfortable — remember, you may have to stay still for several hours. It is remarkable how much wildlife can be seen by sitting quietly in a parked car in a forest car park, a country lane or even a lay-by on a main road.

If you do go looking for animals or their signs, remember that their interests come first. Never disturb the animal, its home or the habitat on which it is dependent. Many species are now strictly protected by law.

KEEPING A RECORD

It is important to keep a record of what you see. If you are going to take this seriously then it is a good idea to carry the following items in a rucksack or shoulder bag in the car.

○ Notebook, ruler, tape, pen and pencil.

○ Mixing bowl and stick to mix plaster.

○ Casting rings.

○ Plaster (in polythene bag). Surgical plaster is best.

○ Water in container.

○ Newspaper.

○ Petroleum jelly.

○ Polythene bags to collect specimens.

○ Envelopes for hair, feathers and small specimens.

○ Solid containers for delicate specimens.

○ Camera.

○ Small tape recorder.

○ Binoculars.

It might not be convenient to have all of this, but the notebook is essential and a great deal can be learned from taking plaster casts.

There is no substitute for a well-kept notebook containing sketches, measurements and careful notes. Use a spiral reporter's notebook in the field and transfer material to a permanent loose leaf folder frequently. The book can be organised around species, types of sign, habitats or just day by day.

The information you record should include records of sightings, careful notes on and measurements of tracks, precise locations, habitats, weather and time. In this way an overall picture of animal activity in a particular area can be built up over a period of years.

MEASURING AND DESCRIBING TRACKS AND TRAILS

Muntjac track
(track with slots)

1 length
2 length of main track (slots)
3 splay of slots
4 inner wall
5 sole
6 outer wall
7 heel
8 dew claws
9 width of track

Registered coypu trail

1 stride
2 straddle
3 hind track
4 fore track
5 tail (accessory) marks
6 median line

Ferret track
(track with pads)

1 length
2 width
3 hind, heel, ankle or proximal pad
4 palm, hand or medial pad
5 toe or distal pad
6 claw marks

PLASTER CASTS

These are a permanent and very accurate means of recording tracks. Before taking a cast, measure and sketch several tracks (if possible) to show any variations. Sketch the trail, noting surface conditions, weather, location and habitat as shown in the example of a coypu trail. Once this is done select a track which displays the most detail. Carefully remove any small stones or sticks which may be obscuring the details. Place a cardboard ring to fit around the track. This should be about 3cm deep and large enough to allow a margin of 1.5cm around the print. The tops of yoghurt cartons, pieces of plastic drain pipe or even tops of plastic flower pots of different sizes make excellent permanent rings, but must be greased inside before any plaster is poured in. Mix enough plaster to a thick-cream consistency to cover the print to a depth of about 2cm. Pour the plaster gently into the mould, avoiding pouring directly on to the actual track. Plaster gives off heat as it dries. The best place to take casts is in mud or silt as sand gives a poor impression and snow melts. Wait until the plaster has hardened, gone hot and cooled down before lifting the cast. Take it gently from the ground, remove the worst of any soil adhering to it, and wrap it in newspaper. At home ease the remaining soil away under running water, but do not use a stiff brush as this will destroy details. Once the cast is finally dry remove the ring and pick out the raised area of the track in black ink or paint. Label the cast by scratching details into the back. When taking a cast do not litter the countryside with bits of plaster.

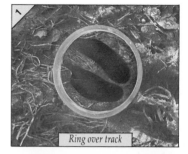

Ring over track

Plaster in ring

Freshly lifted cast

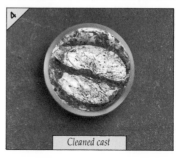

Cleaned cast

PHOTOGRAPHS AND SOUND

Most signs, and the animals themselves, are best recorded by taking photographs. Tracks do not generally photograph well, but homes, remains of feeding activity, pathways and droppings do. It is best to have a camera with a tripod and zoom lens in the 50—200mm range. While colour slides are easy to store, a colour print with a scale bar gives information which can be measured and compared.

With the correct lens — and patience — you can photograph rabbits, fallow deer and many other animals

Prints can also be stored on the loose leaf system along with the notes and drawings for easy reference.

Feeding remains, such as gnawed nuts, cones and woody shoots can be preserved simply by storing in a dry place, although cones hold their shape better if they are wrapped in muslin as they dry. Nuts, gnawed or nibbled by different species, can be stuck on cards and labelled or kept in handily sized old plastic film containers.

Professional sound recording equipment is expensive, but the microphone capabilities on many modern cassette recorders are such that you will be able to pick up a lot of animal sounds, especially if you are prepared to wait patiently through dusk and into darkness for a few hours. Some animals, such as the fox, produce a range of clearly identifiable sounds, but it has to be said that sounds are of limited interest since they tend to be short-lived and are often confused.

MAMMALS

Mammals are warm-blooded, have fur on their bodies and give birth to live young which they suckle and take care of. Several 'Orders' are represented in Britain — the Insectivores (moles, hedgehogs, shrews), Chiroptera (bats), Carnivora (dogs, foxes, cats, weasels), Lagomorpha (rabbits, hares), Rodentia (squirrels, voles, mice, rats, dormice, coypu) and Ungulata (sheep, goats, deer, cattle, pigs). Horses are in the Perissodactyla order.

Most mammals are active throughout the year, although some become less active or even hibernate. There are always signs of mammal activity irrespective of the season or the place.

BADGER

Meles meles

This is a distinctive animal with a black horizontal stripe on the side of its white head. The body is grey on the top and black on the underside.

The tail is stubby and grey. It is the size of a heavily built small dog (body and head length of male up to 80cm, female up to 78cm), the slightly larger male having a broader skull and thicker neck. The snout is pointed and strong, and the legs are short and powerful with long claws on the front feet.

The main factors in the choice of habitat are the availability of cover, well-drained soil which is easy to dig, a plentiful food supply at all times of the year and relative freedom from human disturbance. Deciduous woodlands and copses with pasture and arable land close by are favoured. Hedgerows and scrub areas are popular sites for sets, which are also found in quarries, cliffs, moorland, embankments, natural caves, rubbish tips and, occasionally, under buildings. Low-lying marshy areas are avoided

Playful badger cubs

and little use is made of land above the tree line (600m) in the British Isles.

The badger is widespread over the British Isles, being absent only from the Orkneys, Shetlands, some of the Western Isles and the Isle of Man. The numbers vary, but the species is fairly common in the south-west, parts of Wales, northern counties of England, southern Scotland and the whole of Ireland.

This is a sociable species living in groups. It is active mainly at dusk and night. Regular routes are followed between set, latrines, feeding and watering places. There is a reduction in activity in winter, but no hibernation. The set is a network of underground tunnels and chambers, often with several entrances. This network may be very extensive and there is sometimes more than one set in the area of land occupied by a group. Sleeping chambers are lined with large bundles of straw, bracken, grass and moss bedding, which are collected in large quantities in spring and summer. Discarded bedding is sometimes found on the soil mounds outside entrances,

and it may be 'aired' at certain times of year. Successive generations may occupy and enlarge a set. Mating takes place in spring, but delayed implantation means that the young (up to four) are not born until the following year. The cubs stay below ground for the first eight weeks and remain with the female (sow) for about a year.

The badger is omnivorous, feeding on a wide range of plant matter, including windfall fruit, nuts, berries and underground parts of woodland plants. Grass and clover are eaten in winter and cereals are taken in summer. Animal food includes small mammals, amphibians, all types of grubs and invertebrates as well as carrion (including large birds), especially in winter. Poultry are occasionally taken and larger mammals, such as rabbits, may literally be turned inside out. The most important single source of animal food is earthworms. The badger makes a lot of noise just moving through the undergrowth, digging for grubs and roots and when it is 'slurping' earthworms. It can also be quite vocal, producing a range of sounds from the high-pitched squeals of cubs to the deep warning growl made by an alarmed adult.

Unless you can badger-watch at a set the chances of seeing the animals are limited. They are large, however, and they leave many very obvious signs which are easily seen. The set — with discarded bedding and excavated soil outside actively used holes — and the footpath network are prominent. Near the set, tree trunks and posts may show claw scratches and also marks low down where the animals rub themselves when they emerge. Sycamore trees may have their bark stripped to reach the sweet sap inside. Wasp and bee nests may be torn open if they are near the ground. Snaffles (shallow scrapes to get at roots and grubs) are common on the woodland floor. Fields of cereals can be damaged by badgers feeding on them and rolling them down in play. Where paths cross under bushes or a wire fence, tell-tale silvery grey hairs are a sign of recent activity. The droppings are found in latrine pits and vary greatly in consistency, but have a mild musty odour (as does airing bedding) and often contain beetle wing cases and visible plant remains.

Badgers move with a clumsy, noisy ambling trot with their heads down and the rear end swaying from side to side. The track consists of a large, bean-shaped pad with five toe prints, all with claw marks, arched some distance in front of it. In soft ground a heel mark shows. The front foot (60 x 55mm) is larger than the hind (60 x 50mm) and has much longer claws. When the animal is walking the tracks are nearly registered and turn slightly inwards.

In a walking trail the tracks are 15—20cm apart, but badgers can move quickly and as the pace quickens registration decreases and stride increases. When the animal gallops it leaves groups of four tracks, common in the weasel family, with at least 40cm between each group.

The badger's set is often very extensive

BATS

Bats are the only mammals capable of true flight. In many ways they are similar to other mammals, but the finger bones are greatly extended to carry the skin which forms the flying membrane. The knees bend backwards, so that, with the toes facing backwards also, it is easier for the bat to land and hang upside down. The ears and, in

roost by day in buildings, caves, mines, cellars, tunnels and hollow trees. Bats may move to different roosts at different times of year but tend to use the same ones year after year.

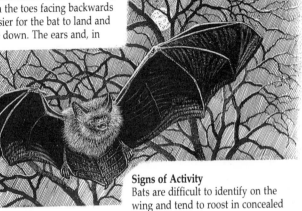

some species, facial skin structures, are modified for the 'echo-location' system which helps bats to navigate and to search for food on the wing. There are two groups of bat found in Britain. Two species belong to the *Rhinolophid* or horseshoe bats. They take their name from the complex skin folds on the face. These bats hang free when resting and will often wrap their wings around their bodies. They cannot move very well on the ground. The other group, the *Vespertilionids*, have no facial skin, generally hang against a surface and move quite well on ground.

Bats do not breed until they are three of four years old, and some species may live for up to 30 years. They all feed on insects in this country and have to hibernate in winter when food is not available. They hibernate for part of the year, and at other times

Signs of Activity

Bats are difficult to identify on the wing and tend to roost in concealed places but they do leave signs. Within roosts and below them droppings are a tell-tale sign. Large accumulations of droppings and sometimes the remains of larger prey, such as beetle wing cases and moth wings, may be found in summer roosts. In winter roosts, odd droppings and small scratch marks on rocks in caves may be the only signs of bat activity. Individual droppings, although very small and on first inspection rather mouse-like, vary in shape and size with the species. Location will also give a clue to species. Droppings are cylindrical, brown to black in colour, porous, crumble easily and consist almost entirely of insect remains. Away from roosts, accumulations of large insect remains may be found below favourite 'perches' where bats hang upside down to consume their prey.

Bats do sometimes land on the ground, either by accident or design. On odd occasions tracks and trails will be found in very soft mud by water or in cowpats. Pipistrelle bats have been recorded walking and running. The serotine walks while the noctule has been reported to run and 'leapfrog'. The tracks consist of a small dot or 'thumb print' from the tip of the folded wing and a normal, small five-toed hand-shaped outline from the hind foot. Because the thigh bone is curved and the knee joint modified, the hind feet point out almost at right angles to the median line. In a walking trail, where the limbs are moved in diagonally opposite pairs, the tracks are close to the median line and the hind almost register over the fore. The body is held close to the ground and there are often accessory marks from the body and tail. When the animal runs, the wings are spread wider so the 'thumb prints' are away from the

median line, the stride increases and accessory marks disappear. Sizes vary considerably. In a pipistrelle walking trail, the hind feet are about 6mm long (with claw marks) x 6mm wide, with the 'thumb print' about 4 x 3mm and a stride of about 30mm.

The hind tracks of serotine bats are about 9 x 9mm, while the 'thumb print' is tiny and the stride is 55mm. The hind tracks of a running noctule bat are 12 x 15mm, the thumb print about 7 x 7mm and the stride 50mm.

Bats and the Law

Bats are at worst harmless and in some parts of the world are very important — in pollinating certain plants, for instance. In Britain, under the Wildlife and Countryside Act, 1981, it is illegal for anyone without a licence to intentionally kill, injure or handle a bat, or to possess a bat whether dead or alive. It is an offence to disturb them in their roosts or to damage these roosts. This applies everywhere except in the actual living space of a house from which the bat may be gently removed. Information from the Nature Conservancy Council.

Bat in walking position (note position of hind foot)

The following bats of the *Vespertilionids* group have a long pointed tragus (flap of skin) in the ear. There is no complex skin fold on the nose and the tail is contained in the tail membrane, or sticks out only slightly.

PIPISTRELLE
Pipistrellus pipistrellus

This is the most common of all bat species in Britain. It is also the smallest of British bats, with the body up to 4.5cm long, the forearm 28—34mm long and the wingspan up to 25cm. It has short rounded ears with a short blunt tragus. There is an extension of the tail membrane on the outside of the bony projection from the ankle. The fur colour varies from medium or red-brown to almost black, but is slightly lighter on the underside.

Pipistrelles are found in all types of habitat, except high, exposed areas, and are often seen flying close to water and over marshy ground. They are frequently associated with human habitation, and are found over the whole of the British Isles, except for the Shetlands.

They fly late into the autumn and may not go into hibernation until December. Bats seen flying in mild spells in winter are normally of this species. Large summer breeding colonies of many hundreds of females may form in roofs. These come together in June and disperse again in August when the young are flying. In summer the males roost singly or in small groups. In winter the sexes hibernate together in large concentrations, mainly in buildings in Britain. The single or, occasionally, twin young are born late in June or early July after delayed implantation from mating the previous autumn. The young remain in the roost until they can fly at about three weeks, and are fed by their mothers. This species roosts in confined spaces, rather than large cavities, and is almost exclusively associated with human activities. Cavity walls, spaces under tiles, and gaps between a porch and a house wall are good examples of sites.

This species feeds on a wide range of insects, smaller ones being taken and eaten on the wing, larger ones eaten at a perch. After hibernation both male and female weigh about 4g while in autumn they are over 6g. In captivity, bats have been recorded consuming over 1.5g of food in one short feeding session with pregnant females eating 3g a day.

BROWN LONG-EARED BAT
Plecotus auritus

This small bat is distinguished
from others by its enormous ears. The
bases of the ears touch each other
on top of the head. The coat is grey/
brown and looks brown when parted.
The tragus is narrow and thin. When
the bat is roosting the ears may be
folded down to protect them. The body
is up to 4.8cm long, the forearm 34—
41mm and the
wingspan up
to 28.5cm.

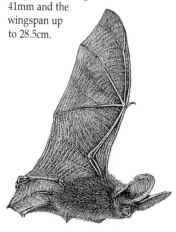

This species prefers wooded
country and quite high densities may
occur locally. In terms of numbers this
ranks second after the pipistrelle. It is
widespread, and fairly common in
suitable habitats, over the whole of the
British Isles, except the extreme north
of Scotland, the Orkneys, Shetlands
and some of the Western Isles.

This species does not generally
emerge until after dark and flies
intermittently through the night, but
returning to the roost before dawn.
The flight is fluttering and rather weak.
It hunts amongst trees, frequently
hovering with the body at a 30 degree
angle to the horizontal to take insects
from the leaves. Insects are also taken
on the wing. Hibernation begins in
mid-November. Roosts in buildings are
mainly in crevices in brickwork,
although roof spaces are used in high
temperature conditions. Sometimes bat
boxes may be used for hibernation.
Trees are important roosting sites and
occasionally caves may be used.

Breeding colonies are established
mainly in the roofs of houses. Females
segregate from males in early June, the
young from the previous year often
staying with the pregnant females.
These bats do not start to breed until
their second year. This is generally a
silent species. Analysis of prey taken
back to roosts shows this species has a
preference for *noctuid* moths (a certain
family of moths) in the summer. As
with other species, signs of activity
consist of droppings, moth wings and
beetle wing cases below feeding roosts.

DAUBENTON'S BAT
Myotis daubentoni

This small bat has a body up to 5cm
long, a forearm of 33—40mm and a
wingspan of up to 27cm. The fur is
reddish-brown on the back and the
muzzle is pinkish. The tail membrane
is slightly hairy. This bat lives in
woody country, although it is also
known as the 'water bat' because it
often flies over water skimming the
surface to capture 'rises'. It favours
slow-moving water rather than fast-
flowing streams.

Summer roosts are in trees and
buildings, but caves and tunnels are
used in winter. Colonies may be large
in summer, but these disperse in the
winter and individuals may hibernate
in a horizontal, rather than hanging

position, in the rubble on cave floors. This species, which is widespread over the British Isles except in the north-west and west of Scotland, emerges just after sunset and often flies regular routes, such as woodland edges.

WHISKERED BAT
Myotis mystacinus
This is a very small bat with dark skin, dark greyish coat and small feet. The body is up to 4.8cm long, the forearm 31—36mm and the wingspan up to 24cm. It is present over much of central and southern England, but records elsewhere are limited and it probably occurs more widely in northern England, Wales and Ireland. It favours wooded country, roosting often in large nursery colonies (ie where the female gives birth to and raises the young), in trees and buildings (especially roof spaces) in summer, and

dispersed in caves in winter. The flight is slow and fluttering. It emerges in early evening and may even fly in the day. The food consists of insects and spiders, which are sometimes taken off foliage. Whiskered bats have been recorded as living for up to 19 years in captivity.

NATTERER'S BAT
Myotis nattereri
This is another small bat. The fur is grey-brown and there is a dense fringe of hairs on the edge of the tail membrane. The tragus is long and there is a slight angle in the hind margin of the ear. The body is up to 5cm long, the forearm 36—43mm and the wingspan up to 30cm. This species is widespread and locally common over much of the British Isles, being absent only from the north of Scotland and the north of Ireland. It is found in open country, with woodland close by, and parkland, and is often associated with

Whiskered bats in flight

water. The bat roosts in trees, buildings and caves and is normally solitary in hibernation. It emerges from roosts shortly after sunset and is active during several periods throughout the night. Hibernation occurs from December to March.

Its food consists of insects, which may be taken on the wing when the bat is flying slowly, or may be picked off foliage as the bat flies along a hedgerow or woodland edge. These may be flicked up with the tail. Natterer's bats have been reported to live for up to 12 years in captivity.

BRANDT'S BAT
Myotis brandti
This species is similar to the whiskered bat in many ways, but has a reddish-brown back fur and a buff underside. The body is up to 4.8cm long, forearm 32—37mm and the wingspan up to 22.5cm. It inhabits buildings in summer and caves in winter (it may hibernate deep in these caves). This bat is generally similar in habits to the whiskered bat. Its range is uncertain as there is sometimes confusion between species, but it is present over much of south and central England.

BECHSTEIN'S BAT
Myotis bechsteini
This is a small to medium-size bat with the body up to 5cm long, the forearm 39—44mm and the wingspan up to 30cm. It has very long, widely separated ears which extend beyond the tip of the nose when folded forward. The coat is reddish coloured. This forest and woodland species is found in parts of central and southern England. It hunts amongst the trees shortly after sunset and flies with a slow wingbeat. The bat roosts in tree holes in small numbers throughout the year, though the species occasionally uses houses and caves.

SEROTINE
Eptesicus serotinus
This is a large, dark-skinned, dull-brown bat with the tail tip projecting beyond the membrane. The ears are rounded. The body is up to 7.5cm long, the forearm is 48—55mm and the wingspan is up to 38cm. It is essentially a woodland species usually roosting in tree holes, although

Caves are used by brandt's bat (below) and the serotine (top, page 29)

In winter, roosts may be in trees or external fissures in walls. This species emerges before dusk and tends to fly high, though it is frequently seen hunting close to water. It takes food on the wing and also from the ground. The droppings are distinctive in that they contain the remains of whole moths, whereas the majority of bats discard the wings.

sometimes buildings will be used in towns and villages and, occasionally, caves. Medium-size breeding colonies may form in summer. Restricted mainly to south-east England, it emerges well before dark, flying straight, but also diving to catch large beetles and moths on the wing. It sometimes lands on foliage in order to catch prey.

The noctule roosts extensively in woodland

NOCTULE
Nyctalus noctula
This large bat has a distinctive uniform golden-brown coat. The body is up to 8.2cm long, the forearm 46—55mm and the wingspan up to 39cm. It is present, but not common, over much of England, Wales and southern Scotland. This bat is mainly a woodland species which roosts, for preference, in mature and over-mature trees. Large noisy colonies may form in trees in summer. Natural or woodpecker holes in trees may be used as roosts. These holes are often stained by the bat's droppings, urine and grease.

LEISLER'S BAT
Nyctalus leisleri
This species is similar to, but smaller than, the noctule and with darker roots in the golden coat. The body is up to 6.4cm, the forearm 39—46mm and the wingspan up to 34cm. Another woodland species, it is present over much of England and Ireland, and roosts mainly in tree holes — often close to water — though it occasionally uses buildings. Similar to the noctule in habits, this species takes to the wing at dusk and catches large insects in the air close to the tree canopy.

BARBASTELLE
Barbastella barbastellus
This is a small, very dark bat with a short muzzle and short ears which meet on the top of the head. This last feature distinguishes it from all other species. The body is up to 5.2cm, the forearm 36—43mm and the wingspan up to 28cm. It occurs locally in England and Wales as far north as Yorkshire, with a preference for woodland, roosting in both trees and buildings, and is sometimes associated with water. These bats tend to be dispersed in summer, roosting in small tree crevices, but often concentrations will occur in caves in winter. The flight is slow, direct and often low over water. Food is taken on the wing.

HORSESHOE BATS
There are two species in Britain. They have a complex skin fold on the nose and they wrap their wings tightly around their body when roosting. They tend to be colonial, often breeding in very high densities, and have a low fluttering flight pattern.

LESSER HORSESHOE BAT
Rhinolophus hipposideros
This is a small bat with the characteristic 'horseshoe' around the nose and a light-coloured coat. The body is up to 3.9cm, the forearm 35—42mm and the wingspan up to 25cm. It is found mainly in south-west and central England and Wales, with secondary populations as far north as Yorkshire, and is also present in Ireland. The numbers are not high. This bat is found mainly in wooded country and roosts in caves, tunnels and cellars in winter with large breeding colonies in attics and buildings in summer.

The large breeding colonies tend to disperse in winter in this species, although a number of individuals may hibernate close together. They emerge shortly after sunset and remain active throughout the night.

GREATER HORSESHOE BAT
Rhinolophus ferrumequinum
This large bat has the typical 'horseshoe', is light brown on its back and has thick wing membranes. The body is up to 6.8cm, the forearm 43—49mm and the wingspan up to 39cm. It is restricted to parts of south-west England and south Wales. This species has habitats and habits similar to those of the lesser horseshoe, forming large breeding colonies in open roof spaces, especially in wooded valleys. It emerges after sunset, taking and eating much prey on the wing, although some is taken to a perch to be eaten.

COYPU

Myocaster coypus

This large <u>aquatic</u> rodent is a native of South America. It has a heavy, square head, small ears, large orange front teeth and a long, cylindrical tail with a thin covering of short hairs. The large hind feet are webbed between four of the five toes. The head and body are up to 60cm long, the tail up to 45cm.

end in a chamber about 60cm diameter. Small platform resting nests may also be constructed in reed beds. Apart from eating occasional freshwater mussels, this is an entirely vegetarian animal.

There are many signs of activity, Signs of digging to reach roots and the extensive tunnel systems are perhaps

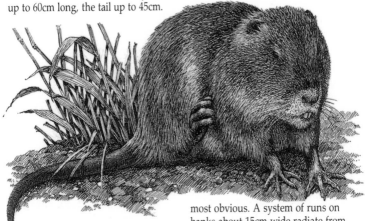

It is found only in the Norfolk Broads and surrounding rivers, in river banks, reed beds and marshy areas with dense aquatic vegetation. It is an excellent swimmer and diver, but is not happy on land. It prefers the stillest parts of rivers, lakes or ponds.

Coypus are mainly active at night and around dawn and dusk, resting in nests in riverside vegetation or burrows in the day. They live in related groups. There are up to two litters consisting of as many as nine coypus, although four is the more usual number. The young may be born in any season, and are able to swim soon after birth. Large breeding nests are normally in burrows which are about 20cm in diameter, up to 6m long and

most obvious. A system of runs on banks about 15cm wide radiate from 'climb out' points on the banks. The droppings are long, cylindrical and spindle-shaped, with fine ridges running down the sides. Deposited on the bank or in the water, they vary from about 7 x 2mm in the very young to 70 x 11mm in adults. The tracks vary greatly in size with age, sex and size of animal. Both fore and hind tracks show five toes. The front tracks are up to 60 x 60mm and unwebbed, the hind are up to 150 x 80mm and are webbed between four toes only. The toe prints are very heavy and show broad, continuous claw marks. There are distinct palm pads on the hand-like tracks. The long tail shows as a continuous wavy line and makes the trail quite unmistakable.

RED DEER
Cervus elaphus

This is a large deer with the head and body up to 200cm and shoulder height up to 120cm in Britain. The rump is buff coloured, not white, and is obvious. The coat is reddish-brown in summer and greyish-brown in winter. The calfs are spotted. During the autumn rutting season the stags develop a long neck mane. Fully mature males develop large antlers which have two points pointing forward from the main stem. In young stags there is only one forward point which forms a right angle with the main stem. Some stags, known as hummels, only develop small knobs instead of antlers. The antlers grow inside a velvet coating from spring to early autumn, lose the velvet by September and are shed in spring.

Red stags roar during the autumn rut

This deer is now most abundant on moorland and upland grassland. The animals may move into woodland or agricultural land off the hills in some areas in winter. The native population occurs mainly in Scotland, north-west England and perhaps south-west Ireland. Elsewhere there are <u>feral</u> populations in parts of Yorkshire, the north Midlands, East Anglia, central Wales, parts of the south-west and the south around the New Forest.

Red deer are most active at dawn and dusk, both on open land or in woodland. On the hills they move upwards to rest in the day before returning to the main grazing areas at night. The woodland groups feed on local fields after lying up in the day. This is generally a sociable species, forming large groups in open country and smaller groups in woodland. The old stags are solitary in winter, and the stags and hinds generally live separately except in the rut. The main rut is in September to October when the stags round up as many hinds as they can defend. The single, occasionally two, young are born the following summer and, after the first few days, accompany their mothers and are suckled by them for up to 10 months. There is no home as such, only flattened lying-up places in undergrowth, moorland vegetation or rough bracken.

Red deer both graze and browse (ie, eat from trees and shrubs). Normally, red deer are fairly silent animals, but the stags 'roar' in the rut and the alarm is often signalled by a series of short barks from the hinds.

Such a large animal leaves many signs of activity. The large, open networks of paths in woodlands are particularly obvious, as are browsing lines and bark gnawed to a height of nearly 3m. The droppings, about 2 x 1.5cm, are dark when fresh, but fade quickly. They are often adhesive and deposited in groups. Individual 'currants' are elliptical in shape in females, pointed at one end and flattened at the other in males. During the pre-rut summer period, the males produce miniature cow-pat like droppings up to 7.5cm across.

Wallowing places, where males coat themselves in mud, and rubbed branches and tree trunks, where males have been removing velvet from their antlers, are again very clear signs. The lying-up places in dense vegetation can belong to various species, and droppings are the best guide. Shed hair is another useful indicator. Antlers are shed in February to March and are often nibbled to reclaim the nutrients so they are not a common find. The tracks are large, up to 80 x 70mm in the males without <u>dew claws.</u> Females leave smaller prints and those of young animals are relatively longer and more pointed, but there are common distinguishing features. There is a distinct outer wall on the cleaves and the toe pad is large. The cleaves do not tend to splay widely and the very small dew claws rarely show. At a walking gait the tracks are imperfectly registered and tend to be straddled. In the trotting trail the tracks are almost heel to toe and in a straight line. In the galloping trail the tracks are grouped in fours, normally splayed and with 2—3m between each group.

Velvet antlers on the red stag in summer

FALLOW DEER
Cervus dama

This is a medium-sized deer with a body up to 170cm long and standing up to 110cm at the shoulder. There is much variation in the background coat colour, ranging from almost white to nearly black. In summer the most

common colour is reddish-fawn with white spots on the flanks and back. There is a black stripe down the spine. Some dark forms show virtually no spots. The white rump, edged with black, and the long tail with a black mark down it are distinctive. The young are spotted with a reddish-brown coat. The antlers of mature males are distinctly flattened into the broad, flat palmate form which also distinguishes this species.

Wild and escaped herds are found in broad-leaved, coniferous and mixed woodlands with dense undergrowth and also in open woodland and parkland. In undisturbed conditions the deer may come close to human

habitation. There are many captive herds in parks. This species was re-introduced into Britain in the Middle Ages. It is widespread over England and Ireland with limited populations in Scotland. The wild herds are the descendants of the original stock which inhabited the Cannock, Epping, New and Pickering Forests.

This is a <u>gregarious</u> species which sometimes forms into large herds, especially in parkland. There are well-established territories. Much of the day is spent lying up and resting, but this deer is active over 24 hours and is likely to be seen in undisturbed areas at any time.

Fallow deer feed throughout the day, but mostly at night. They graze a great deal, especially on grasses and sedges, and will take the new leaves from broad-leaved trees and a wide range or fruits and nuts, with acorns and beech mast being particularly popular. Bramble, ivy, holly and the foliage and bark of felled conifer trees are important food sources in winter.

In the wild, herds are loosely organised with female and male groups sometimes together and at other times apart. The rut occurs in October to November and the single fawns are born the following June. The young remain hidden in the foliage to avoid detection in the first few days of their lives, but are following their mothers by the time they are two weeks old. This is a species which is silent for much of the year. In the rut the male makes a loud belching noise called 'groaning'. The females may bark if

Fallow deer give birth to a single fawn

frightened, and the young bleat. If the animals are disturbed, feet may be stamped as a warning to others.

There are many indications of fallow activity. The antlers are shed in May and are more likely to be found than many other deer antlers because of their large size. Trees and the soil are scented with a pungent smell in the rut. Rutting areas are marked by scrapes (30—300cm across) in leaf litter, grass or soil. Rutting rings around trees are not commonly made by this species. The rutting scrapes are urinated in by the bucks, who fray young trees and bushes with their antlers and plaster themselves with urine-soaked mud in the rut. The signs and smell of this activity are obvious. The tracks, up to 65 x 40mm, are heavy, with the inner walls parallel. The outer walls are often concave, a distinguishing feature in this deer. Slipped or deeply impressed tracks show the small dew claws. The animal walks, runs, jumps and gallops. If alarmed it may first move off with the peculiar stiff-legged bouncing action known as 'pronking' before breaking into a run or gallop. Although able to jump well a fallow deer will go through a hedge, low vegetation or even a damaged fence rather than jump them. In a walk the tracks are about 60cm apart, in a gallop they are in groups of four and up to 110cm apart. The droppings are pellets found in small piles or in a string. Black in colour (16 x 11mm in males, 15 x 8mm in females) and sometimes with flat sides, they are pointed at one end and concave at the other. Lying-up places in bracken or coarse grass normally have a cow-like smell, but this may be rather acrid at certain times of year.

ROE DEER

Capreolus capreolus

This is the smallest native British deer with the head and body up to 120cm long and standing up to 75cm at the shoulder. It has virtually no tail. In summer the coat is reddish-brown and the rump patch is pale and not obvious. In winter the hair changes to a greyish brown and the rump is white and more obvious. The fawns are spotted. The antlers are short and have a rough ring around the base and not more than three points. Antlers are fully developed in May and are shed in early winter.

All kinds of woodland are inhabited, especially those with clearings. The deer often feed along woodland edges and in neighbouring fields. In Scotland, roe may be found on the open heather moorland. Large Forestry Commission plantations have done much to increase the populations in northern England. This species is widespread over Scotland and the northern counties with secondary concentrations in Hampshire, Dorset and also Norfolk, but is absent from Ireland. There are also introduced populations in other parts of the country. This species is increasing its range in many areas.

The males are territorial all through the year. Roe deer are usually solitary or in small groups of male, female and young, but may form into small herds in winter, normally still segregated by sex or age. They generally remain in cover in the day, but are active throughout the day in undisturbed areas. There are frequent fights between bucks. The males mark territory by fraying trees and gouging with their antlers. If alarmed the deer bound away barking and with the rump flared. If they are disturbed, but not alarmed, they slip quietly away. Rutting takes place from mid-July to mid-August, fawns are born between April and July, and twins are common. The young lie up for the first few days of life and then follow their mother. The deer have no permanant homes, but lying-up places are made in dense undergrowth. Both the males and females bark. Males bark and make rasping noises as part of their territorial behaviour. Females utter a high-pitched squeak in the mating season. The fawns squeak or bleat.

This is mainly a browsing species, feeding from trees and shrubs, but it will also graze. Bramble is a major food source all year round. Ash, hazel and oak are important in summer. Grasses, herbs and conifers are also crucial in some months. On the hills, heather, bilberry, hazel twigs, spruce and grasses are taken. Roe deer feed for 8—10 hours in the 24-hour cycle, sometimes from dusk to dawn, but if undisturbed during the day as well. They are often seen moving through clearings or feeding on the edge of tree blocks in large plantation areas.

There are many distinctive signs, including small trees 'frayed' and bark rubbed by antlers until wood is exposed in the rutting season. Males mark territory by scraping soil with the feet and also annointing vegetation with scent. Well-trodden circular paths are made around bushes, trees or tree stumps and are used by males, females and young in July and August but their significance is not known. Bark is stripped to a height of 100cm on young conifers. The droppings are 15—20mm long, usually pointed at one end and depressed on the other, and are sometimes deposited in regular latrines. Lying-up places have a slightly musty odour. The tracks are delicate, up to 50 x 40mm, and pointed towards the front. The inner walls run outwards at the back of the track. The tips are deep and the dew claws commonly show. Front tracks in particular are widely splayed and often appear uneven. At the walking pace tracks are in a single line, in a gallop they are in groups of four and up to 2m apart. Paths through dense undergrowth and unthinned conifer trees are clear to a height of 1m.

This small native deer has short antlers

CHINESE WATER DEER
Hydropotes inermis

This is a small deer without antlers. The coat is reddish-brown and there is no obvious rump patch. The ears are large and the males have large tusks. It is found in marshes, grassland and open woodland. An introduced and

escaped species, it is restricted to south and east England as far north as Norfolk. It is a solitary and nocturnal animal which grazes on root crops and grass. It barks and makes a screaming noise when alarmed, and is thought to be territorial. The males make a whistling noise in the rut, which takes place in December, and the young are born the following May or June. There may be as many as four young.

Activity signs include well-marked open, narrow pathways in reed beds, and droppings which are small, black and cylindrical, and may be deposited in small groups or in regular latrines in lying-up areas. The currants are 5—15mm long and may form a <u>crottie.</u> The tracks, relatively large (30—50mm) and evenly impressed with slightly convex inner walls, are one way to identify the species. In widely splayed or slipped tracks a web may show between the cleaves and dew claws are often present. Even in the walking trail the tracks are slightly straddled rather than being in a straight line.

MUNTJAC
Muntiacus reevesi

This is a small deer, standing no more than 48cm at the shoulder, with a glossy chestnut-coloured coat. The light-coloured rump can only be seen when the tail is raised. The male has tiny backward-pointing antlers with two small points. Antlers are lost in May or June and regrown by November. The permanent antler <u>pedicle</u> is a distinguishing feature. The preferred habitat is broadleaved woodland with thick undergrowth. This species is found mainly in southern and eastern England where it has colonised from escapes, but may be extending its range. Active by day and night, it is most commonly seen

around dusk. It browses on ivy, bramble and tree seedlings, but also grazes and will take fruits. Muntjac can produce a very loud bark, as well as clicks, squawks and a very loud distress call. Many signs are left in inhabited woodland. Pathways are open up to 60cm above the ground, and there are browsed bushes and lying-up places, frequently containing large quantities of the small black droppings, either as individual currants or as crotties. The track is the most distinctive of all deer. It is small (under 30mm long), pointed and narrow with the two cleaves always offset giving an asymmetrical appearance. The normal gait is a walk, but this deer can run very well and it is also an extremely agile jumper if under pressure.

SIKA DEER
Cervus nippon

This medium-sized deer (up to 150cm long and 75cm at the shoulder) has a chestnut-red coat with white spots in summer and greyish in winter with fewer spots. The white, heart-shaped rump patch with a dark area at the top is distinctive. The tail is short and white, sometimes with a slight grey streak. The antlers never have more than two forward points. Sika deer prefer woodland which is deciduous or mixed and with dense undergrowth, though they are also found in the early stages of conifer plantations. Feral herds are found in Dorset, Hampshire and Yorkshire, Argyll and Caithness in Scotland, and Dublin, Kerry and Wicklow in Ireland. The social organisation and feeding habits of this species are similar to those of red deer. The rut is mainly in October, with one male tending to dominate an area.

A sika stag

Single young are born in May or June. Sika produce many noises including whines, whistles and grunting roars. This animal is most active at dawn and dusk, moving into the open to graze, but retreating to cover in daylight.

It is difficult to tell if bark has been stripped or shoots browsed by this deer rather than by another sort. Stags create 'rutting platforms'. The droppings are black and similar in shape to, but smaller than, those of red deer, and are often deposited in regular latrines. The tracks are up to 80 x 50mm with indistinct toe pads and slightly concave inner walls. The dew claws are close together and set high on the limbs, so they rarely show.

Sika deer's track

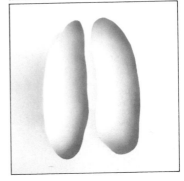

HAZEL OR COMMON DORMOUSE
Muscardinus avellanarius

This attractive little orange-brown creature has a long, bushy tail. The head and body are up to 90mm long and the tail up to 75mm. The ears are short and the eyes large. The common dormouse is found in and around deciduous woods — especially those with nut-bearing species such as beech and hazel — with dense undergrowth.

It is most common in the south-west, but recent evidence suggests it may occur as far north as Yorkshire. This is a largely nocturnal animal. There are many distinctive signs of activity. The summer nests are often in shrubs and are made of stripped honeysuckle bark, leaves, grass and moss and do not have an obvious entrance. Hibernating nests are often at, or below, ground level in stone walls, under tree roots or even in burrows. Nest boxes may be used in all seasons. Hazel nuts are gnawed in such a way that the hole is uniquely rounded and polished.

Although little time is spent on the ground, the hind feet with their small, clawless thumb leave a very distinct track (15 x 11mm) while the forefoot is a normal four-toed structure.

EDIBLE OR FAT DORMOUSE
Glis glis

This large mouse (body up to 190mm, tail up to 150mm) is rather like a small squirrel. It has dark eye rings, and has been recorded in mature broadleaved woodlands, orchards, gardens and houses. Edible doormice live in several colonies around the Tring area north-west of London. They are active mainly by night, feeding heavily to fatten up before hibernating. The summer nests may be in the canopy of a tree. Nuts are split, leaving large jagged holes, willow and plum trees may have strips of bark torn off and the remains of fruit, such as apples, may be stored in roof spaces. In common with squirrels, this dormouse may leave its feeding remains on tree stumps. The hand-like five-toed hind and four-toed fore tracks show small, rounded almost clawless pads and the tracks turn slightly outwards. The bushy tail drag is commonly seen between the widely straddled tracks.

THE BADGER

This distinctive animal is widespread throughout nearly all of Britain. Its main periods of activity are at dusk and night.

The badger emerging from its set

A registered badger track

The set — a network of tunnels and chambers — may have a number of entrances

Badgers use regular routes, creating pathways through the undergrowth

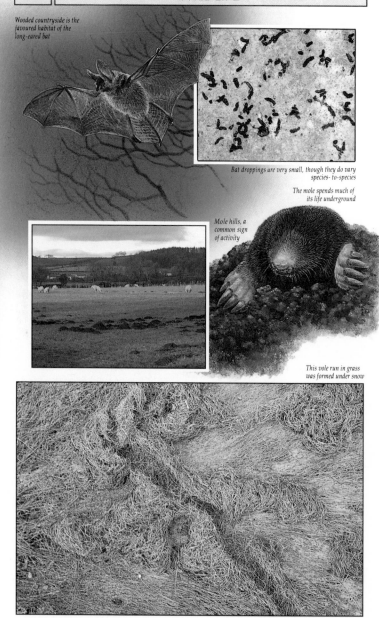

Wooded countryside is the favoured habitat of the long-eared bat

Bat droppings are very small, though they do vary species- to-species

The mole spends much of its life underground

Mole hills, a common sign of activity

This vole run in grass was formed under snow

Rook

Jackdaw

Crow

Included for comparison with the droppings of mammals (see page 84 for selection) are the pellets of three common birds

Bats sometimes live in specially constructed 'bat boxes'

The roe deer, the smallest native British deer, has short antlers and a tiny tail

The significance of the circular paths or 'rutting rings' created around trees by roe deer is not completely understood

A small tree 'frayed' by roe deer

Antlers from roe deer (top) and fallow deer

FERRET

Mustela furo

The dark forms of this medium-sized member of the weasel family may be similar to polecats, with a pattern of white on the head and a light underside. It is generally much lighter and may even be albino. The body is up to 44cm long, the tail up to 18cm. Wild populations (which will breed with polecats) live in a variety of conditions, seeming to favour open land with cover nearby. This species was widely kept in captivity for the control of rabbits and there have been many escapes over the years.
There are

This is a totally carnivorous species eating rabbits, hare, mice, voles, hedgehogs, amphibians, reptiles and perhaps invertebrates — as does the polecat. It kills with a bite to the neck. This is a vocal animal. It will scream when angry or frightened and both young and old animals make chattering noises when contented.

also well-established feral populations in places like the Isle of Man, Anglesey and parts of Yorkshire.

Feral animals are solitary and mainly nocturnal. They spend most of their time on the ground where, when they are moving at speed, the back is arched to give a sinuous gait. Ferrets normally live in holes in the ground, including burrows of other animals. They may shelter in farm buildings as well. A single litter of between 5 and 10 young is born in May to June.

The activity signs may be confused with those of the polecat. The stink or musk glands below the tail may be used to mark territory, or emptied in fear or anger. The droppings are 60—70mm long and about 5mm wide. They are dark, normally contain fragments of bone and fur, are twisted within themselves, but not coiled and are tapered at both ends. They have a strong musty smell. The tracks are five-toed with widely spread toe prints set around distinctly lobed palm pads. The hind tracks are about 45 x 40mm and the fore 35 x 35mm. Occasionally, tail marks may be present. The tracks are very similar to those of polecats, but the palm pads tend to be heavier.

FOX

Vulpes vulpes

The fox is very distinctive, with a narrow muzzle, large ears and a long pointed bushy tail with a white tip. The body is up to 75cm long and the tail, or brush, is up to 45cm. The coat is usually a deep reddish-brown with a light underside, although there is considerable range in colour and there may be dark markings on the neck and chest. The overall appearance is

sometimes speckled because of the long black/yellow/brown guard hairs which form the outer coat. The front of the forelegs is almost always dark.

This is traditionally a woodland animal, but farmland, moorland and even mountains are inhabited. The fox is a very adaptable animal which has had its range extended by coniferous plantations, especially when these are newly planted, and increasingly over the last 20 years by its ability to survive in built-up areas. It occurs all over the British Isles, except the Isle of Man and some of the Western Isles. Because of its versatility in habitat, activity patterns and feeding, it is often seen, as well as being apparent through its signs. Sometimes the fox is solitary. Young males without

territories tend to move about a lot, but territory-holding males may have a single mate, or several perhaps closely related ones. The territories range from 10—100 hectares. Although active mainly at dusk and in the night, the fox is often seen in the day, especially in undisturbed conditions. Undisturbed does not mean quiet — it refers in this context to places where human access is difficult, since foxes are often seen in busy railway sidings or 'dead ground' beside motorways. Adults generally spend the day lying up in dense vegetation, in short tunnels the fox has dug itself, in a burrow made by another animal or under sheds or some other form of human-provided shelter, such as dry drainage or sewer pipes. Foxes like the sun and may sometimes be startled from sheltered coarse grass or field edge areas where they doze on a warm day. Both young and adults spend much time in play. They are strong swimmers. Although this is a territorial animal there is a great deal of overlap between groups if plenty of food is available.

The family groups may stay loosely together, although the vixen (female) generally looks after the cubs. A single litter of four or five young is born in late March. The vixen stays with cubs constantly for the first three weeks of their life and then rests away from them in the day. The home, or earth, may be a purpose-dug burrow, a hole shared with or taken over from another species (such as a badger), in rock piles, or in buildings and a large range of man-made features. There may be a number of earths in the territory and the animals move between these. The earths can be distinguished by the acrid smell,

Fox cubs are born in early spring

feeding remains and droppings outside. This is a highly vocal animal, often for several hours after sunset. Sounds include intermittent high-pitched barks and a hoarse wailing bark. In the breeding season an ear-piercing screeching, interspersed with chattering, is common.

The fox will eat most things. Rabbits, hares, voles, other rodents, birds (especially ground-nesting species), beetles, earthworms (which are very important in some environments), eggs, carrion and refuse are all taken. The fox only takes small numbers of lambs, but can be a real problem with poultry, waterfowl and game birds, such as pheasants and grouse. Fruit and berries are important elements in the diet in autumn.

Many signs tell of the fox. The scent deposited in the urine is very strong and persistent, and is often the most obvious sign, wafting considerable distances on the breeze. Urine and anal gland secretions are used to mark territories by both sexes. Any suitable point may be marked, including grass tufts, stones and tree stumps.

Kill remains are crushed across the neck and back. Feathers are ripped out and the large primary wing feathers are sheared cleanly through near the base. Bones are crushed on larger prey, but eggs have the end neatly bitten off and the contents

Adaptable or cunning? The fox survives in many habitats

lapped out. Feeding remains and droppings are common outside earths, and partially buried food, such as half-eaten carcasses of larger kills and even eggs, are a common fox sign. These remains may be hidden over a large area and are sometimes covered with vegetation rather than being buried.

Regular pathways are followed. Runs through hedges are smaller and narrower than those of badgers, but larger than those of hares or rabbits. Pathways in dense vegetation are obvious because of the distinctive reddish hair caught on thorns or wire. Droppings are often deposited on prominent places, such as stones, molehills and small tree stumps, to mark territories. The droppings ('castings') are twisted with tapered ends and are often linked together by hair or feathers into chains up to 20cm long. They are generally dark when fresh, but bleach with age and contain bone, feather and fur fragments. They may be confused with owl pellets, but close examination will show the droppings to be twisted with fragments parallel to the sides. (In owl pellets the fur/feathers and bone fragments are randomly mixed and there is no twisting. Fresh owl pellets also have a moist coating.) The tracks are delicate and four-toed, the palm pad being the same size as one toe pad. The forefoot is larger than the hind (tracks about 50 x 45mm). In a trotting trail the tracks are evenly spread and almost in a straight line, in stark contrast to the erratic path of the domestic dog, where the paw pads are also much larger than the toe pads. Foxes will stalk, run, bound and are known to make small bounds when they are playing.

DOMESTIC DOG
Canis familiaris

The dog has been included here, out of alphabetical sequence, so that its signs may be compared with those of the fox. There are many different breeds of dog and, although some may look rather fox-like, they originate from wild dogs and wolves which are a different sub-group completely to the foxes within the *Canidae* (dog family).

The alsatian, a popular pet

Dog signs include tracks, which are four-toed, have a large palm pad and vary greatly in size with breed. The tracks are straddled and at an angle, reflecting the peculiar 'crab walk' where the hind quarters are carried to one side of the fore. This contrasts markedly with the fox, where the palm pads are very small and the tracks always tend to be straight, purposeful and unstraddled.

The only other dog signs are scratchings in grass and soil, urination marks and droppings. The latter are sausage-shaped and, although colour, size and contents are variable, they are neither twisted or coiled as they are in the fox. The droppings are also left at random by dogs, whereas the fox uses prominent marking places or urinating points, mainly to mark out territory.

GOAT (*feral*)
Capra hircus

This primitive breed is derived from escaped domestic goats. Mainly long-haired and often piebald, they are quite small in build and both sexes carry horns. These are small in females, but may be very large in males. A characteristic feature of the horns in the wild strains is that they turn outwards and upwards and not simply backwards, because they have an additional spiral twist. Some animals have no horns at all.

Feral goats are found in rocky areas and moorland from sea level to over 300m. There are goats in Scotland, the Scottish Borders, Wales and Ireland. There may be a dominant male, but other males also compete for mates in the breeding season. The herds are active throughout the day, grazing and browsing on almost all forms of vegetation. Outside the breeding season the sexes tend to remain separate, although males may be seen with females and kids at any time of year. There is no harem formation, and the one kid, or occasionally two, will be born in late winter or spring. There are no permanent homes, but lying-up depressions are established behind walls, boulders and in small wooded areas in bad weather.

Although they look different from domestic breeds, many of the signs left by feral goats are similar. Regular paths often follow the contours on the hillside. These may be confused with those created by sheep, but can be distinguished by the droppings and tracks. The droppings are cylindrical, dark, flattened at the ends and non-adhesive. They are about 10mm long and are dropped in groups rather than strings. The tracks are up to 60 x 50mm and are curved and rounded at the ends. The cleaves splay open at the front, but touch at the back which is a distinguishing feature in goats. The dew claws are high on the leg and rarely show. The tracks lie side by side rather than being registered. Although the goats tend to move slowly, they can move at speed and galloping trails show the tracks in groups of four, widely splayed and pointing outwards.

BROWN HARE

Lepus capensis

Golden-eyed and long-legged, this is the largest member of the rabbit and hare family. The body is up to 65cm long, the long hind feet up to 150mm, and the distinguishing, very long, black-tipped ears are up to 105mm in length. The coat is yellowish-brown and the black and white tail is again a distinctive feature.

This species is often associated with permanent pasture, but also with cereal areas next to pasture on farmland. It is sometimes present in open woodland and on the moors, although it is not common above 500m. The brown hare is rarely found more than 1km ($\frac{1}{2}$—$\frac{3}{4}$ml) away from farmland and is not especially associated with long vegetation and dense cover.

It is widespread in England and Wales on lower ground. In mainland Scotland it is present on farmland and rough grazing into the far north-west, but is absent from some of the Western Isles and Ireland generally, where it has been introduced but is not spreading widely. Populations vary and numbers are not as great in many highly improved agricultural areas as formerly. The highest densities occur on downs and along plains in the south and west.

While this is mainly a solitary animal which is most active at dusk and by night, it is sometimes seen in the day, especially in quiet areas. This is particularly true in spring when the pre-mating 'boxing matches' and chases take place between a male and female. Small groups may be formed. Hares will sometimes bound and run away very quickly if disturbed, but will also crouch with ears flattened back unless approached too closely.

Brown hares breed and shelter in shallow surface scrapes known as 'forms'. There may be several of these in a territory. The young, normally two, are born fully furred and able to see. These 'leverets' are camouflaged and remain still in shallow scrapes and depressions in vegetation for protection in the first few days of life, but then move around with the female who will leave them lying up and concealed by surrounding vegetation while she feeds or rests.

Farmland is the brown hare's favoured habitat

Food consists of grass in spring and summer, when crops are also taken. Root crops may be eaten in winter and bark is stripped on a limited scale. Young woody shoots are clipped off at the tips. This is normally a silent animal, but it will scream if hurt or alarmed and sometimes makes low grunting noises.

Various signs of hare activity can be seen. Shallow depressions excavated against cover, such as boulders or stones, are used for shelter or breeding. Breeding nests may be lined with vegetation. The droppings (12.5—15mm diameter) are circular, slightly flattened, paler, more fibrous and slightly larger than those of rabbits. They are moist and adhesive when fresh, becoming bleached and brittle with age. They will be deposited in piles in shallow scrapes or at random over the territory.

Frayed bark stripped from the base of trees shows distinct, double front teeth marks, as with the rabbit and mountain hare. Small twigs and young trees or bushes which have been eaten by a hare show a clear-cut, angled break. Moulted hair, yellowish grey with black roots, is sometimes found in the forms. The paths across cultivated land are distinctive and are characterised by depressed areas where the continuous landing of the large back feet gives an irregular effect to the path surface.

The tracks are obvious. The large slipper-shaped hind foot (up to 150 x 40mm) always shows four toes, sometimes with a fifth toe and a complete paw outline in very soft ground or snow. The much smaller forefoot (40 x 40mm) has only four toes. Neither have palm pads, although hair will often show between the toes. On hard ground only the extreme tips of toes may show and these will give a very confusing track. In a gentle hopping trail the hind feet appear side by side with the fore feet behind each other in between. The stride is about 25cm. In a bounding trail the tracks are widely splayed in groups of four with up to 250cm between each group.

MOUNTAIN OR BLUE HARE
Lepus timidus

Slightly smaller than the brown hare (up to 60cm), this animal has notably shorter ears (up to 80mm) which are black tipped. The tail is all white. In summer the coat is greyish-brown, turning white in winter except for the ear tips. In Ireland the coat remains a lighter yellowish-brown all year.

It is found in a range of upland habitats, including grassland, heath, moorland and open woodland, and may come into arable land and pastures at lower levels where the brown hare is absent. The main populations are found in the Scottish Highlands and Ireland. It is also present in the Isle of Man, a small area of north Wales and the Peak District.

Feeding generally takes place at night, but there is much activity in the day, especially in less-disturbed areas. If frightened, the leverets (young hares) will bolt into burrows but the adults tend to run away. The animals establish long, narrow territories up and down the hillside.

The form (a shallow surface shelter) tends to be towards the top of the land occupied by an animal with the feeding ground lower down. Unlike the brown hare, this species may dig short burrows (about 2m long) or even take over old rabbit warrens. There may be up to three litters a year. The behaviour pattern of the young is the same as in brown hare.

Short, well-managed heather is the preferred food, accounting for about 90 per cent of the total in winter and 50 per cent in the summer. 'Resting' forms may be burrows or a deep trench in snow, vegetation or peat. In the spring these will often contain downy, white moulted fur as the animals spend a great deal of time resting up in open areas in the day. The droppings are small (10mm diameter), fibrous, grey-green or brown in colour, and circular. They are left at random on the hillside or in small scrapes. The tracks are slightly smaller than those of the brown hare (hind up to 130 x 30mm, fore 35 x 30mm), and a dense covering of hair over the foot in winter may distort them. Otherwise the structure of both track and trail is similar to that of the brown hare.

HEDGEHOG

Erinaceus europaeus

The hedgehog is unmistakable with its coat of over 5,000 spines which are 22—25mm long, its long pointed snout and uniformly coloured underside which ranges in tone from dark brown to beige. The compact head and body are up to 275mm long, while the underside is hairy.

This species is common where there is grassland close to woodland, scrub or hedges — woodland edges, hedgerows in pasture and even sand dunes with shrub on them are preferred habitats. It is present in almost all habitats where there is cover for nesting, including gardens in the country and suburban areas. The hedgehog is found below the tree line all over mainland Britain and the islands, with the exception of parts of the Western Isles and Orkneys.

This is an almost entirely nocturnal animal, although it is sometimes seen at dawn and dusk and often as a road casualty at certain times of year. Hibernation takes place in purpose-built nests from about October to April, but this varies with age, sex and place. It is generally solitary, although several animals may have overlapping territories. Hedgehogs tend to follow regular routes, some of which may be well marked. Perhaps the best known habit of the hedgehog is that of rolling into a ball when danger threatens, therby exposing the spines.

There may be several nests in the territory. Hibernating and breeding nests are made from dry leaves and grasses, hibernating nests being carefully made and hidden under brambles, in rabbit burrows, compost heaps or under garden sheds. More than one nest is used as animals move around in mild spells in winter — do not destroy the nest in the compost heap in January, as the owner may come back. In summer, non-breeding adults may shelter in dense vegetation.

The young are born in the purpose-built nests, and they may be eaten by the female if she is disturbed soon after giving birth. There may be two litters of four or five babies between June and September. Young remain in the nest for three weeks, but often stay with the female into autumn.

Hedgehogs feed almost entirely on insects and other invertebrates living in and on the ground. They are also known to eat small numbers of bird chicks and eggs, as well as carrion and some small mammals (which may already be dead). Hedgehogs sometimes kill amphibians, but the frequency with which adders form part of the diet is not as great as folk tales suggest. Certainly hedgehogs are not immune to the venom, but they are well protected by their spines. When alarmed hedgehogs squeal. They also make a lot of noise crashing through the undergrowth, snuffling and snorting as they go.

There are many signs of activity, including the nests and small areas of ground dug over for grubs and worms. Horse droppings, dung heaps, compost heaps and rubbish will all be snuffled over in search of insects. Cow pats will be broken open and are a good place to look for tracks. Tracks are five-toed (hind 45 x 25mm, fore 40 x 25mm). The claw marks are long and continuous with the deeply impressed toes. Hand outlines are common and the broad heel distinguishes the tracks from those of rat and squirrel. When moving slowly the hedgehog shuffles along with feet flat on the ground, but turning outwards and straddled. Spine marks often show in soft soil. When the animal runs, less of the foot shows and spines are absent as the body is clear of the ground. The droppings, a common and obvious sign on the lawn, are cylindrical (up to 15cm long), slightly twisted and with little smell. They are hard, compressed and often black coloured, containing the dark remains of insect wing and body casing. If larger animals have been eaten the droppings are lighter and fragments of hard parts like bone, feather or fur may be seen.

The hedgehog is usually seen at night

HORSE AND PONY
Equus caballus

There are a number of fully domesticated breeds of horse and pony. In many parts of the country, with increasing interest in riding, the signs of horse activity are more common than they were 20 years ago.

It is worth mentioning the Domestic Donkey, *Equus asinus*. This is a separate species to the horses and ponies, although its signs are similar. Originally, the donkey was derived

Welsh mountain pony and foal

from the wild ass of North Africa, and it has never been native to Europe as a wild species. In southern Europe there are large numbers of donkey, some living semi-wild, but elsewhere numbers are small and the species is totally domesticated. In Britain, it is now largely confined to seaside pleasure beaches, though there are considerable numbers still in use in agriculture in Ireland.

The animal illustrated here belongs to one of the various groups of semi-wild ponies found on Dartmoor, Exmoor, the Lake District, the Welsh Uplands, Northumberland, the Shetland Islands, Western Isles and south-west Ireland. All of these animals are dependent on people to some extent, although they spend much of the time fending for themselves. It is thought that some of these breeds may be descended from original wild horses which lived in the British Isles some 7—10,000 years ago, but most originate either from escapes or introductions from later domestic stock.

Semi-wild herds of ponies are active mainly by day. No specific shelter is sought although lying-up places may be made in sheltered areas. This is mainly a grazing species, although a wide range of fruits and bark are also taken. The voice needs little description. Single foals are born and remain with their mothers, often for more than a year.

Shetland ponies in their native habitat

Horses, ponies and donkeys have limbs which are completely specialised for running at high speed. The bone structure has changed so that the animals now stand on the extreme tip of the third toe, the finger and wrist bones being greatly elongated and the other toe bones very much reduced in size or completely absent. The resulting tracks, although these will vary in size with age and breed of animal involved, are very distinctive. They consist of a single large imprint with flat sides and a blunt, rounded front, with a deep notch mark in the back. In hard ground often only the outer rim will show and in fully domesticated animals only the shoe outline may be present. Tracks of a small pony may be only 10 x 10cm,

while those of a shire horse may well be 25 x 25cm. The distances between the tracks will again vary with breed and size, but when horses walk the tracks are straddled across the median line and appear in groups of two. At the gallop tracks register poorly, or not at all.

Droppings, which consist of large yellow-green masses of unchanged cellulose from partially digested grass stems or hay, are the most obvious sign of this species. Horses in fields will gnaw at their fences and both wild and domestic varieties will strip large areas of bark from trees by chewing with their front teeth. Ponies will also rasp areas of bark from the lower parts of tree and shrub trunks.

MINK (*American*)
Mustela vison

The true wild form is dark, but there are many variations. All minks are uniformly coloured, except for a white area on the lower jaw. This is a medium-sized member of the weasel family with a shortish bushy tail (head and body up to 40cm, tail up to 14cm)

This species is found close to rivers, lakes and marshy areas. It originally colonised from numerous fur farm escapes, but is now present on most river systems in the British Isles.

Much time is spent close to and in water. The mink has partially webbed feet and is an excellent swimmer. It is active mainly at night, through the year.

A wide range of food is taken, including waterside birds, water voles, rabbits, other small mammals, salmon and trout. Activity signs include the dens, with their distinctive musty smell. The droppings, not deposited in regular places, are twisted and coiled, loosely constructed, covered in mucus when fresh and very strong smelling. They contain the remains of scales, bone and shell and are much smaller (20—30mm) and stronger smelling than otter. Remains of large kills have bite marks at the base of the skull.

Although mainly a nocturnal species, mink can sometimes be seen in the day in quiet areas

Riverside dens, and some dens away from water, are made in tree roots, tree trunks, holes and crevices among stones. There are often droppings outside and a strong musty smell is present. The young are born below ground. The 2—6 young are active with the mother from June onwards, but disperse in late summer.

Tracks are frequently found by the waterside. Fore and hind feet are five-toed, with a web between the toes. The small tracks (hind 45 x 35mm, fore 30 x 40mm) have long slender claws. There is a large palm pad, and a small heel pad sometimes shows in very soft conditions. On land the mink moves with the typical arched-back gallop of weasels. It travels along the river bank, sometimes on land, sometimes in water. Tail drag marks are very common in soft conditions. While the tracks are of similar size to those of the polecat and large stoat, the much larger palm pad and well-developed web on the mink track are useful clues.

MOLE

Talpa europaea

This small animal (head and body up to 150mm) with a short tail has a beautiful fur of black velvet. The front feet are large, flattened and stick out sideways. The eyes are tiny, but can be opened and the long tapering snout is covered in sensitive whiskers. There are no external ears.

The mole's complex tunnel system with hill and 'fortress'

The mole is abundant in woods with good ground cover, in grassland close to woodland, scrub or hedges, in gardens, parks and playing fields, though it can also be found in some damper meadows and dune slacks. This animal is widespread on mainland Britain, but is absent from Ireland, the Isle of Man and some of the Western Isles.

Much of its life is spent below ground in tunnel systems, which are dug by alternate strokes of the powerful front feet, the loose soil pushed behind the body and periodically up vertical tunnels, again with the front feet, to form mole hills. Moles are active throughout 24 hours, and mainly dependent on the sense of touch through the highly sensitive

snout. They are able swimmers, and can sometimes be seen on the surface, especially in times of drought.

Males make long, straight tunnels in spring to cross female systems. The tunnel systems are really food collecting areas and the true nest is often in or below a much larger mound of earth or so-called 'fortress' than in the surrounding mole hills. Sometimes, it may even be a simple nest. Close to the nests there are deep permanent tunnels, further out they may be more shallow. Sometimes tunnels close to the surface collapse and under snow shallow runs, half in soil, half in snow, may be formed and these are seen after melt. Shallow surface runs, known as 'traces d'amour' are associated with breeding activity in spring.

Moles produce quiet twittering noises when excited and single noisy squeaks when fighting — aggression is common if animals stray into each other's tunnel systems. The food consists mainly of earthworms, although larger soil insects are important. Stored food consists mainly of earthworms with the head segments chewed.

The most obvious activity signs to look for are the mole hills, runs and fortresses. The number of mole hills and runs is related to the length of tunnels and the heaviness of the soil. In light soils there are less hills. Tracks are occasionally found on bare soil. The mole walks on the side of its highly developed front feet. Only the five toe tips show in an 'L' shape and are about 15 x 10mm. The hind feet look like small hands. In the rare trail which may be found, the tracks are widely straddled and the body drag marks are obvious.

HOUSE MOUSE

Mus musculus

The eyes, ears and hind feet of this mouse are small. Its tail is relatively thick and obviously ringed, there is no light chest spot, and the coat is generally brownish-grey above but may be much lighter. This mouse has a characteristic notch in the upper front teeth. The body is up to 95mm, the tail up to 95mm and the hind foot up to 19mm. This species is often associated with houses, farm buildings, warehouses and industrial buildings in towns. It is found in gardens in rural areas and as a hedgerow species in the country where other small mammal species are absent. It may be present on arable land, and also on cliff habitats and disturbed ground. In such suitable habitats, it is widespread over the whole of the British Isles.

Bulky nests are made of shredded grass in extensive burrows in hedgerows, under stones or other debris. In hay stores, where optimum conditions exist, communal nests may be made. In buildings the nests are made of paper, cloth or similar soft material in wall cavities, roof spaces or under floor boards. This species breeds throughout the year if food is abundant. There are five or six young in a litter and a litter may be produced every three or four weeks.

This is a highly territorial species in areas with high population densities. In open country, the house mouse tends to be solitary and

nocturnal, but is more sociable and active throughout the day when associated with human habitation. It is active all year, and in the countryside moves from open land to buildings during autumn and winter. This species creates regular runs and tunnels.

In built-up areas all forms of human food are raided. Much damage may be done to stored food, more by urine and dropping contamination than by actual consumption. Cereals are the main food source, with green foods and fruit of secondary importance. Insect food is sometimes taken and even plaster, soap and some plastics will be nibbled. In open land, cereals, grass and weed seeds are eaten.

There are numerous signs of house mouse activity. These include gnawed and spoiled food, runways, smear marks, holes, droppings and tracks. Holes have a strong 'stale' smell. Half-eaten grain in stores will show teeth marks, or the husk will be powdered. Droppings are concentrated in specific places. They are small (6—7mm long, 2—2.5mm wide) and the colour varies with content. In buildings the droppings may be associated, on runs, with urinating pillars, which are accumulations of droppings, dirt, grease and urine in habitual stopping places. Dirty black body smears may be present along frequently used runways. Regular pathways in dust show tracks, smears and finely chopped food. The tracks are slender, the hind print up to 18mm long x 12—18mm wide, the fore track about 10 x 13mm. There are five toes on the hind foot, four on the forefoot. The wavy line of the long tail frequently shows in the trails.

YELLOW-NECKED MOUSE
Apodemus flavicollis

This is similar to the closely related wood mouse, but slightly larger (body up to 120mm, tail up to 135mm, hind foot up to 26mm). However, it is often brighter in colour and has a distinct yellow chest spot, which sometimes forms a collar.

It is found mainly in woods, hedgerows and field edges. Often associated with human habitation, its greatest densities are found in south-east England and the Welsh border counties. There are also records as far north as Yorkshire and it may also be present in the south-west.

Activity patterns and signs are generally similar to those of the wood mouse. However, the yellow-necked seems to be a more adept climber, appears to do less damage to crops and has a slightly larger track (hind 24 x 19mm, fore 16 x 18mm).

The mountain hare's coat turns white in winter

One of the many ponies of the New Forest

The water vole is most often seen at dawn and dusk

The hazel or common dormouse lives amongst deciduous woodlands, especially nut-bearing trees

Right: *The pipistrelle bat can walk and run*

Below right: *The coypu can be seen on river banks in the Norfolk Broads, but is not happy on land*

Below: *The bank vole, with its distinctive russet-coloured back, is active by day and night*

The red deer in classic pose amongst highland moor and mountain

Tree stumps are a favourite feeding place with the wood mouse

Fallow deer may be seen grazing at day, though they mostly feed at night

The fox, most active by night, usually spends the day lying up in thick vegetation

A grey squirrel drey in a birch tree

The dominant grey squirrel is now far more common than the red

Right: *Trail left by the hopping rabbit*

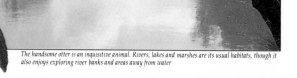

The handsome otter is an inquisitive animal. Rivers, lakes and marshes are its usual habitats, though it also enjoys exploring river banks and areas away from water

WOOD MOUSE

Apodemus sylvaticus

This is one of the larger mice with big eyes, ears and feet. The upper body is yellowish-brown with a pale silvery underside and small yellow neck mark. The body is up to 110mm, the tail up to 115mm and the hind feet up to 24mm.

This is the dominant small rodent species in woodland, also occurring in hedgerows, scrub, cliff habitats, gardens and often buildings. It is the most common species of mouse and is widespread over the entire British Isles. Wood mice are either solitary or live in small groups. They are mainly nocturnal, but have peaks of busyness at dawn and dusk (especially in winter), and are active all year.

The home is a nest of grass and leaves, which is thickened in the winter, in an underground tunnel or crevices under tree roots, logs or stones. Females raise the young. Litters of 4—7 are born throughout the year if there is an adequate food supply. Food consists of a wide range of nuts, fruits, buds, seedlings, snails, insects and earthworms, though the species will also eat bark and fungi. The wood mouse often has regular feeding places on tree stumps.

Among the many tell-tale signs are acorns and hazel nuts gnawed to leave an unpolished hole with corrugated edges and teeth marks around the hole. Rose hip seeds are eaten and the flesh discarded. Grain is nibbled, leaving coarse dust. Snail shells may be tackled by biting through them away from the spiral. Shallow tunnels established in the soil and litter layers of the woodland floor may last for many years. Droppings are 4—8mm long, rounded in section, short and thick. They are deposited in prominent points, which are also feeding places. Droppings are initially pale, soft and moist but darken, dry and harden very quickly. The tracks are relatively slender (hind 22 x 18mm, fore 13 x 15mm). Complete hand outlines are common. The stride ranges from about 4cm in the walking trail to over 50cm if the animal leaps. Tail drag is not seen in the trails.

This mouse has large eyes and ears

HARVEST MOUSE

Micromys minutus

This is the smallest European mouse with small ears, an attractive rich orange brown coat and a very long, slender, slightly <u>prehensile</u> tail. The body is up to 75mm long, the tail up to 70mm.

This mouse is associated with tall ground cover and long grass areas, such as field edges, river banks, marshes, road verges and scrubland, especially bramble. It occurs in cereal crops, although less than it used to, and in reed beds. Ungrazed meadows which are cut only once a year are an important habitat. It may occur on the edge, or in clearings, of woodland and plantations. Harvest mice have a widespread, although patchy, distribution in England as far north as Yorkshire, with limited records from further north and in Wales.

It is a very active animal, spending much time climbing in the stems. The harvest mouse tends to be nocturnal in summer, but is active throughout 24 hours in winter though it is generally less busy during this season. This species also tends to confine itself to a small area for much of the year. Its hearing is acute.

Nests are close to the ground. During winter nests are at ground level or even below it. Early spring nests are built in hedges. The summer nest is unique. It is built in the stalks of growing cereals, grasses or reeds. The leaves from the stalks are split lengthways into ribbons and woven together without being detached. This well-camouflaged nest grows up from the ground in line with vegetation growth. The nest is 8—10cm diameter with at least one obvious entrance

hole. The breeding season runs from May until late into the year. The litter consists of up to eight young. The blind young make audible squeaks if disturbed.

The food consists of insects, fruits, seeds and berries. Grass shoots are eaten in early spring, grain from cereal heads in late summer.

The most obvious signs, the nests, are often not seen until autumn. Gnawed grain heads show sickle-shaped remains. Food stores may be found close to winter nests. The droppings are tiny, about 2—3mm long, dark coloured and normally only found in and around nests. The tiny tracks and trail are rare (hind 13 x 10mm, fore 8 x 8mm). There are long claw marks, and sometimes a thin wavy tail drag line will show. The stride ranges from about 40—80mm.

OTTER

Lutra lutra

This beautiful animal has the long tapering body of the weasel family, but is characterised by its long thick tail, which gradually tapers to a point. All four feet are webbed, the coat is light brown when dry and there is a light bib on the chest. The body is up to 80cm, the tail up to 45cm.

Rivers, streams, lakes and marshes are the traditional habitats. In Scotland there are many semi-marine otters living on, or close to, the coast. Otters will also travel long distances overland. Away from the coast, rivers, streams, ponds and lakes where there is plenty of ground cover and extensive tree root systems are favoured. The otter is now drastically reduced in numbers, though it is still relatively common in parts of north and west Scotland, parts of Wales and the south-west, and Ireland.

Otters are active mainly at night, lying up in 'holts' under tree roots, rocks, reed beds and even man-made features, such as the piers in Shetland. A great deal of time is spent in the water, where they are very much at home. Otters are a very inquisitive and intelligent species spending much time exploring surroundings close to and away from water.

The holt is not a purpose-dug hole. It is often under tree roots, log jams (or man-made log piles), boulders and in reed beds in marshy land. Old badger sets, fox earths and even rabbit burrows may be taken over in well-wooded areas, often some distance from water. There may be a number of holts in one animal's territory.

Cubs share the female's territory for about a year. A female with cubs is avoided by other otters. The young, normally two or three, are born at any time of year, but most often in spring.

A variety of fish is eaten. Otters also consume crustaceans (crabs and crayfish), aquatic insects,

amphibians and small numbers of birds and animals.

Adults make a short chirp or long piercing whistle. When they are uncertain, a drawn out moan or low growl is made. Otters are rarely seen, but they leave a wide range of signs. Food remains are often found in regular places, such as a fallen tree or waterside boulder. Remains consist of fish carcasses with the fleshy parts eaten away, shells and bones. Regular pathways and travelling routes are followed in undisturbed areas. These consist of obvious bankside paths leading in and out of the water. Sometimes steep 'slides' of polished soil into the water may be made by animals at play. Flattened areas in vegetation in 'rolling places' at the end of short paths from the water are made where otters dry themselves. Bundles of vegetation and long grass stems may be used to build a 'couch', usually found in quiet places. 'Twists' of vegetation are made from sedges, ferns and rhododendrons with the stems laid parallel.

The droppings, or 'spraints', may be deposited in the water or in regular sprainting places, such as a boulder, tree stump or other prominent place close to a run and often near water. Spraints are black, covered in mucus and very musty when fresh, although they weather very quickly. They are coiled and cigar shaped, between 30 and 100mm long and are often found in groups. They contain fish bones, scales, pieces of shell and sometimes bone, fur and feather fragments. Sprainting activity is common on boundaries of territories where population densities are high. Scratching areas and marking spots (such as stones, tree trunks, soil mounds and tufts of grass) are marked by both urine and scent, and are found at feeding and sprainting places. The holts are most often under tree roots, such as oak, sycamore or elm where there is plenty of cover. Close to the sea they may be in a cliff crevice or pile of stones. The track is very distinctive. Both fore (65 x 60mm) and hind (85 x 60mm) have five rounded toe pads with short, blunt toe marks. The web between the toes is often deeply impressed. The palm pads are large and in soft ground a heel pad may show, along with body drag and tail marks.

The otter has regular feeding places

PINE MARTEN
Martes martes

This species is similar in size to a cat (body up to 55cm, tail up to 27cm), but with the short legs and ears of the weasel family, a bushy tail and a long muzzle. The pine marten's coat is reddish-brown, and there is a large throat patch which varies between pale yellow to a light orange in colour.

The versatile pine marten is agile on ground, in trees and water

The marten is largely associated with coniferous or mixed woodland, but is also found in deciduous woodland and will live in open hill areas. Forestry Commission plantations have helped this species in many areas, and it is widespread over Ireland and north-west Scotland, with isolated records in Cumbria, Northumberland, Yorkshire, Derbyshire and north Wales. It is not clear if these smaller populations mean that the pine marten is spreading.

This is a solitary animal, hunting mainly at night or at dusk and dawn. A great deal of time is spent on the ground, but this is an extremely agile climber and can move through the trees at speed as well as being a strong swimmer. Regular runs are often followed. The marten is normally silent, but when angered produces a high-pitched scream though it may produce growls, moans, chattering, purring and clucking noises at any time. Young make a call similar to that of a snipe.

The den is often in a rock crevice or hollow tree. Sometimes an old squirrel drey or crow's nest may be used as a resting place. Mating takes place in late summer. A single litter of about three young are born in the following spring. The young remain with the female until autumn.

The diet is variable, but is mainly of rodents, including red squirrels and small birds. Tits, wrens, treecreepers and field voles are important species. Other foods include carrion, beetles, grubs, bird's eggs and fish. Berries may be important in autumn.

Regular runs, latrines and feeding places on vantage points are useful signs. Feeding remains consist of bones and fur, and are often found near droppings (often on a hill top). 'Scats', with a musty odour, are coiled, dark and slimy when fresh, 4—12cm long and often left in prominent places. A distinctive feature of marten droppings is that they almost always have a proportion of vegetable matter in them. The tracks are large for the body size (hind 85 x 60mm, fore 80 x 60mm). There are five rounded toes and a large palm pad. A single heel pad may be present as may fur marks, and there is sometimes a faint suggestion of a small webbing between toes.

POLECAT
Mustela putorius

This medium-sized weasel is dark brown, except for white on the snout and on the face between the eye and ear. The palish yellow underfur shows through the dark guard hairs, which give the coat a shine in some places and a patchy appearance. Its body is up to 44cm, its tail up to 18cm.

and moss. The polecat will take up residence in farm buildings, old rabbit warrens or rocky crevices. The female cares for the single litter of between five and ten, born in late spring.

This is a true carnivore which will feed on hare, rabbit, mouse, vole, hedgehog, bird, lizard, frog, grub and earthworm. Feeding remains consist of bones, fur and frog intestines, ovaries

The polecat is most frequent on lower ground in woodland, marshy areas and river banks, coastal dune systems, plantations and farmland. Some animals are now adapting to living on the edge of towns. It is currently restricted to Wales and the Welsh border counties, though it is becoming more common in certain areas and perhaps extending its range.

This solitary species, generally most active at night, is also frequently active by day. This activity occurs throughout the year. It rarely climbs or swims through choice. Well-developed stink glands are used to mark territory and musk is also released when the animal is alarmed. It normally moves at a walk with the head and body close to the ground, but when moving at speed shows the arched back movement so common in the weasels.

The nest is normally made close to the ground and is lined with grass

and spawn, often with droppings close by. The droppings, which may be found in outbuildings and are sometimes in large groups, are black and very musty when fresh. Droppings are slightly coiled, twisted and with tapering ends. They are up to about 7cm long and contain fur and small bone fragments, but never fish scales as in mink. The tracks are five-toed and have slender palm pads with long attached claws. Walking and bounding trails, with the tracks in groups of four, are common

RABBIT

Oryctolagus cuniculus

Rabbits are smaller than hares (head and body up to 45cm), with shorter (up to 7cm) untipped ears and shorter back legs (hind feet up to 9cm). They are generally greyish-brown in appearance, but other colours, including black, are not uncommon. The white bob tail is a common sight when the animals retreat to the safety of their burrows.

They are usually associated with grassy areas, dry heaths and scrubland — often close to a woodland edge. Their extensive warrens are made under cover of hedgerows, stream and river banks or other steep slopes. In dense populations warrens are made in woodland, but rarely under conifers. Rabbits favour areas where digging is easy, such as sands, shales or soft limestones, and are abundant on soft sea cliffs and sand dune systems.

They occur over the whole British Isles, although are now generally much less common than in the days before myxomatosis was introduced in 1953. The disease still occurs and wipes out pockets of rabbits from time to time.

This is a highly colonial animal which creates communal burrow systems. Most active at night, rabbits are frequently seen at dawn and dusk and occasionally in the day. They are regarded as serious pests because of grazing activity and damage to root crops and young trees by stripping bark and clipping off small branches. When disturbed they will run or bound away quickly to get underground. They are generally a silent animal, but will scream when attacked. The hind feet are thumped on the ground as a sign of warning.

Their home consists of a series of underground tunnels. Where large numbers of rabbits live in the same area these tunnel systems, or warrens, may be very extensive with many entrances. Entrances are located in steep banks or under dense vegetation, such as bramble. The holes are generally circular and not much more than 20cm across. The breeding nests, made of grass and moss and lined with the female's belly fur, are in separate short breeding tunnels known as 'stops' which are blocked up with soil to protect the young when the doe (female) is away. Litters of 3—7 are

born in spring and summer and a breeding doe may have several litters. Sometimes superficial 'forms' (surface shelters), similar to but less trodden down than those of the hares, are made by the rabbit.

A wide range of plant material is eaten. Many agricultural crops are taken, including cereals, roots, pastures, vegetables and some fruits,

Look out for rabbits at dawn or dusk

as well as young trees. Heavy rabbit grazing keeps turf short — rabbits can change heather heath to a grass turf — but if the grazing pressure is removed then close-cropped areas will change to taller grasses and shrub. Wild grasses are also eaten. Paths lead to holes in hedges and often burrow entrances. In addition to the regular runs there are many signs of digging and scratching activity and short-cropped grass. Fur found on fences and branches is brown, grey and white. Shallow scrapes made for shelter may sometimes contain grey underfur. Gnawed, rather than stripped bark low on trees is a sign of rabbit activity. The branches of bushes and low trees will be nibbled away to produce a browse line 50—60cm above the ground. Double front teeth marks show clearly on bark, shoots and fungi. There are

often tightly grazed areas around burrow entrances, which can be identified from the small fibrous spherical droppings outside them. The droppings, which are about 10mm in diameter, are generally smaller than those of hare and without any flattened sides, as in sheep or lambs. They are sometimes deposited on soil heaps, mole hills, in shallow scrapes and at random. Droppings are black-green when fresh, but weather over time. There is often no smell, but droppings and latrines may be marked with strong-smelling glandular scents to mark territory. Urine patches in snow are a common sign.

The hind tracks are large slipper outlines (60 x 25mm), the fore tracks smaller and square (35 x 25mm). Generally only four toes show in all tracks. There are no palm pads. In the hopping or bounding trail the hind tracks show next to each other, while the fore are placed in a line one in front of the other.

BROWN RAT
Rattus norvegicus

This large member of the mouse family has a long, thick, scaly, almost naked tail and large ears. The body is usually brown on top and grey below, although colours may be darker and even black, and is up to 26cm long, with the tail up to 23cm and the hind foot up to 45mm.

Although mostly associated with human structures, it is often found in open countryside where it nests in field margins, woodland, walls and river banks. This is particularly true where these have been disturbed in some way. It is widely distributed over the British Isles.

Brown rats are most active at night, but are often seen in daylight hours. They are competent climbers and very adept swimmers, often being found close to water. Extensive burrow and run systems are established in territories. Rats have a good sense of smell, which is important for communication between individuals, and the hearing is very acute, but their eyesight is poor.

Burrows are made in banks, ditches or under cover, such as stones, floorboards, logs or tree roots. Entrances are up to 90mm diameter, those dug into soil often with mounds of soil outside them. Colonies often develop from a single pair. Up to five litters with as many as eight 'kittens' in each may be produced every year.

This species concentrates on cereals, but will eat almost anything. In towns rats take all sorts of waste and will even gnaw soap, plastic and wax.

Signs of activity include pathways and runs. In vegetation the runs are narrow, up to 10cm wide, and continuous (by contrast rabbit runs are discontinuous reflecting the hopping, rather than running, gait of the animal). In buildings runs show as dark greasy smears on wood or brick. Gnawing marks are obvious. Droppings vary in colour with diet. They are between 12 and 20mm long and up to 6mm diameter, tapering to points at both ends and may be deposited in latrines or at random. The hind track is hand shaped, with five toes (33 x 28mm) and a long heel. The fore track is star shaped with four toes (18 x 25mm).

The rat with its new born

COMMON SEAL
Phoca vitulina

This is a small seal (head and body up to 1.9m) with both male and female of similar size. Like all seals, it has small, largely inactive, hind flippers and strong front flippers on the sides of the body. In this species the muzzle is short and the profile of the head is concave between the forehead and muzzle. When a seal is seen face-on, the nostrils form a 'V'. When dry the coat is a light mottled colour, and the pups are normally dark coated.

same sandbanks or rocks every day at low tide. Breeding takes place in midsummer, with the single pup being born on a sandbank or tidal rock.

The diet consists mainly of fish, including flounders, whiting and herring. The signs are limited. There is a characteristic scent (not as strong as that of the grey seal), given off by the oil deposits from the body on sand. Droppings are normally passed in water, but are occasionally found on

Seals do not use any type of shelter, but form into large groups with 'watchout' bulls to guard the colony

The common seal favours shallow coastal waters, river estuaries and sheltered bays. This is the seal most often seen on sandbanks and mudflats, and it has been recorded right around the coast of Britain. Concentrations are found in the Wash, Humber, Tees, Firths of Forth, Tay, Dornock and Moray, with small groups all down the west coast of Scotland and the east coast of Northern Ireland.

It spends a great deal of time in the water, but is mainly a sedentary animal on land, hauling out on to the

sandbanks. They are irregularly rounded and like a dog's, 2—3cm diameter, brown or grey in colour, with the remains of fish bones and shell fragments. Seals move across the ground with forelimbs in contact with the surface, the body being dragged between and the hind flippers held clear. The resulting trail contains five-toed tracks with well-developed claws. The toes are parallel to the direction of movement. The track size varies, and is up to 75cm wide in adults. The fore tracks show as evenly spaced pairs, with the body drag apparent. The fore tracks are 30—50cm apart when the common seal is slowly 'hitching' along.

GREY OR ATLANTIC SEAL
Halichoerus grypus

In this large seal the male is up to 3.2m long and the female up to 2.5m. The muzzle is long and the line between the muzzle and forehead is straight or convex. The nostrils are widely separated. The dry coat shows light blotches on a dark background in males and dark blotches on a light background in females. The cubs have a white coat.

This seal spends most of its life at sea. Breeding colonies are established mostly on small rocky islands and it rarely comes on to sandbanks. It has been recorded from almost every part of the British coast. There are many breeding points from the Isles of Scilly northwards, around Scotland and south to the Wash and the north Norfolk coast.

Breeding and non-breeding 'haul-out' areas are normally separate. Large concentrations of animals may occur on hauling-out points. There is social organisation only in the breeding colonies, and the species is less sedentary than the common seal.

No shelter of any kind is used. The largest breeding groups occur on offshore islands, such as the Farnes. The females often return to the same breeding site. The single pup is born at various times of year, ranging from September to October in the north to March to May in the south-west. Bulls acquire territory into which the cows congregate. The pups bleat when they are hungry and hiss or snarl when frightened. The cows snarl or hiss and the bulls make similar, but much deeper noises.

The tracks and trail are similar to, but larger than those of common seal. They are also much less frequent. The oil from the body carries a pungent, lingering smell. Sometimes traces of hair will be found on regular hauling-out places. The droppings are similar in shape to those of common seal, but are larger (up to 4.5cm diameter) and are often light coloured.

Unlike the common seal, the grey rarely hauls out on to sandbanks, preferring rocky shores

COMMON SHREW

Sorex araneus

Like the other shrews this is a small animal with a long, pointed snout. It is recognised by its three-coloured coat with a distinct intermediate colour between the dark-brown back and pale underside. Young animals are paler and have a tufted tail, but the adult tail is naked. The teeth are tipped red. The head and body are up to 85mm long, the tail up to 47mm, and hind feet up to 13mm.

The shrew's sensitive whiskers are used to sense food

The common shrew is found in all habitats where there is sufficient ground cover, such as woodlands, hedgerows, heaths, dunes and mountain scree. It is widespread over the whole of mainland Britain, but absent from Ireland, the Isle of Man, some of the Western Isles, Orkneys and Shetlands.

It is active by day and night and throughout the year. The activity is in bursts, followed by periods of rest. It rarely leaves ground cover and tends to be solitary except in the breeding season. Females with young are occasionally seen 'caravanning' — moving rapidly across the ground in a line holding on to each other by the base of the tail. Extensive, very narrow runways are made in litter and upper soil layers. Shrews are good swimmers.

The nest is made under logs, stones, flat objects (eg sheets of corrugated iron) or grass tussocks. The female may have five litters of up to eight young in a year. Food consists of earthworms and beetles, though insects, spiders, centipedes, woodlice, snails and slugs are also taken.

Signs include tunnels, which can be recognised by their small diameter and flattened cross section. Insect food may be stored in these tunnels. Droppings are very small (up to 8mm long and 2mm diameter) and cannot be used to identify the type of shrew which left them. They are elongated and contain hard parts of insects, which makes them dark. They are left at random in runways. The tiny (hind 10 x 10mm, fore 10 x 9mm) five-toed tracks are found only in very soft material. The tracks are straddled and nearly continuous tail drag is common.

PYGMY SHREW

Sorex minutus

This is a tiny shrew with a two-coloured coat. The back is brown coloured, whatever the age of the animal. The tail is longer and thicker than in the common shrew (body up to 60mm, tail up to 40mm, hind foot up to 12mm).

It occurs in almost every habitat type, but is less common in dense cover, such as woodland, than the common shrew. The pygmy shrew is able to survive in more open areas, such as moorland and mountain conditions where there is relatively little cover. This is related to the availability of the small insect food on which this species is able to survive. It is present over the whole of the mainland British Isles and all islands except the Shetlands. Numbers in a given area are generally lower than those of common shrews.

Habits are very similar to those of the common shrew, except that movements tend to be even more rapid. This species is relatively more active in the day. It does not make tunnels in litter or soil, but will use existing runways made by other animals. It ranges over a larger area than the common shrew and can climb well. The pygmy shrew is probably able to swim as it is often found on wet moors.

The nest is made under various types of cover, including tussocks and stones. Breeding takes place between April and August, when several litters of 4—7 young are produced. Caravanning behaviour (see common shrew entry) is again sometimes seen.

Generally, it is less vocal than the common shrew, but will squeak on occasions. The food is in many ways similar to that of the common shrew, but smaller woodlice, spiders and beetles are preferred. The ability to survive on smaller, surface-dwelling prey explains why this shrew is able to live in more exposed conditions. Tracks and trail are again similar to common shrew, but are smaller (tracks about 5 x 5mm), with a stride no more than 4cm. The droppings are similar and signs are generally not distinguishable. As with the common shrew, remains are frequently found in owl pellets.

WATER SHREW
Neomys fodiens

This is the largest of the British shrews (body up to 90mm, tail up to 70mm, hind foot up to 20mm). It is black on top with a light-coloured underside. There is a fringe of long hair on the whole length of the underside of the tail and on the sides of the hind feet.

Extensive burrow systems are made through soil and litter with the nose and forefeet. These are flattened in cross-section and have entrances above and below water. The shrew pushes through these to squeeze water from the fur. Breeding nests are made at the end of the tunnels. Breeding takes place between April and September, with two or more litters of 3—8 young being produced.

All sorts of invertebrates, small fish and amphibians are taken,

This shrew is most often associated with clean, unpolluted rivers, streams and ponds where there is ground cover. Its distribution is patchy, with concentrations in suitable habitats, followed by large areas where the shrew is apparently absent. It does not breed in Ireland, on the Isle of Man, some of the Western Isles, the Orkneys and Shetlands.

This is a solitary species active throughout the 24 hours, with peaks at night. It spends much of its time diving, swimming and grooming its fur. The saliva contains venom which acts on small fish and even frogs. Away from water, regular tunnels and runways are made as by other species of shrew.

normally from behind. Food may be stored, and stores or regular feeding places with snail shells and partially eaten frogs or fish are sometimes found. The riverside burrows and runs in waterside vegetation are another useful sign. Droppings, dark and often containing insect remains, are deposited at random. Tracks are five-toed and small (hind 14 x 10mm, fore 12 x 10mm) and hand outlines are common in very soft silt, where even the fringe of hairs on the hind feet may show. The trail will show a run or bound with the tracks splayed, pointing slightly outwards and with a stride of about 5cm.

GREY SQUIRREL

Sciurus carolinensis

This squirrel is mainly grey throughout the year, but often with a brownish tinge to the fur. Very dark ones may be confused with the red squirrel, but there are no ear tufts on the grey, which is also larger (body up to 30cm, tail up to 24cm). The long bushy tail has white fringes.

Originally mainly a broad-leaved forest animal, the grey squirrel has adapted to conifer plantations and the food they provide. It is often found in parks and large gardens where there are trees. This is now the dominant species over England as far north as Yorkshire. It is present in the Central Lowlands of Scotland and an area in the north-east of the Irish Republic, but is absent from all offshore islands.

The grey squirrel is a very able climber, spending large parts of the day in the tree tops but also spending much more time foraging on the ground than the red squirrel. It is active throughout the day, peaking in the first few hours after dawn, and is able to swim.

The nest may be in a tree hole but is more usually a 'drey'. Dreys, compact and rounded, are built close to the trunk or main branches. The outer structure consists of hardwood twigs

Its drey is easily spotted amongst winter's bare branches

and large leaves, the inner nest is made from dry leaves, moss and other soft material. In the summer temporary dreys, resembling platforms, are made from hardwood and conifer twigs, especially oak. Cavities in trees will be lined with leaves and moss. Dreys are between 2 and 15m above the ground and are the size of a large football. There are two main periods of breeding activity, in January to March and May to July. Litters are from 1—7 in size. The male takes no part in raising the young.

The usual call is a harsh chatter, often associated with a rhythmic flicking of the tail. Teeth chattering and moaning are also common. A wide range of food is eaten. Oak is a major food source, when available, as are

beech, sweet chestnut and hazel. Much time is spent feeding on the ground in autumn and among the tree canopies in spring.

Activity signs are generally similar in both species of squirrel, although those of the grey are more extensive. Stripped and gnawed bark, nuts opened by a single split after a small hole has been made at one end, and stripped pine cones are all common signs. Surplus beech masts, acorns, hazel and chestnuts may be stored by burying in the ground, lodging in tree crevices, small clefts and actually in the dreys.

Teeth marks in fungi, deep wounds inflicted in bark to reach the sap underneath, feeding 'stumps' with piles of nut shells, cones, hips and berries and similar feeding remains below branches are again obvious signs. The grey squirrel may gnaw bark on tree trunks and scent with urine, which gives rise to dark, strong-smelling patches.

Damage to trees by stripping bark from any part of the tree and taking out shoots makes this a serious pest in forestry, especially to beech and sycamore. Teeth may be sharpened on bone or stone, and scratch marks in bark and on the woodland floor where the animal is climbing or digging are common. The droppings, dark in colour and cylindrical (up to 10mm) or rounded (up to 8mm) are widely scattered on the forest floor. The tracks are similar to those of the red squirrel, but a little larger and heavier. The five toes on the hind, four on the fore and large palm pads are distinctive. The tail is held high as the animal hops or runs on the ground, but may press in the ground if it takes off

RED SQUIRREL
Sciurus vulgaris

The red squirrel is an exceptionally attractive animal, prettier and more graceful that its grey relation. It is red in summer, greyer in winter but always with very obvious ear tufts. Dark, almost black animals sometimes occur. The long, bushy tail fades in colour after the autumn moult. The body is up to 25cm, the tail up to 20cm.

The red squirrel is most abundant in large blocks of conifer trees (where the grey has not replaced it), especially

native scots pine. It is sometimes found in mixed woodlands, hardwoods (especially beech) and more open wooded areas. This species has declined over this century as the grey has advanced. It is still present over much of mainland Scotland, parts of northern England, upland Wales, much of Ireland, parts of East Anglia, the Isle of Wight and the extreme south-west of Cornwall. *(contd pg 85)*

OPENING A HAZELNUT
Several mammals (and birds) eat hazelnuts.
They open them in different ways.

Dormouse

Wood mouse

Bank vole

Water vole

Squirrel

Nuthatch

OPENING A CONIFER CONE
Animals (including birds) have their
different methods of opening conifer
cones to reach the seeds within.

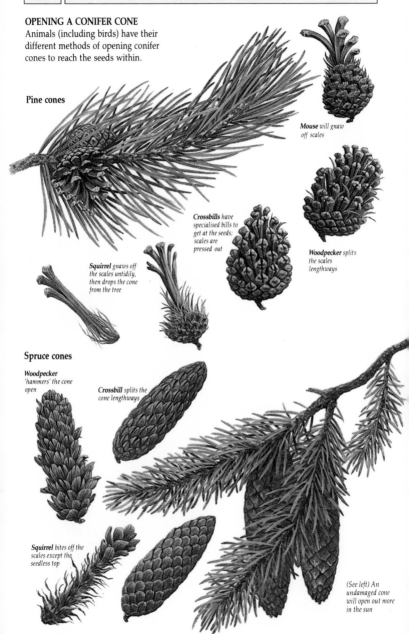

Pine cones

Mouse will gnaw
off scales

Crossbills have
specialised bills to
get at the seeds;
scales are
pressed out

Woodpecker splits
the scales
lengthways

Squirrel gnaws off
the scales untidily,
then drops the cone
from the tree

Spruce cones

Woodpecker
'hammers' the cone
open

Crossbill splits the
cone lengthways

Squirrel bites off the
scales except the
seedless top

*(See left) An
undamaged cone
will open out more
in the sun*

GNAWING OF BARK

Several animals cause considerable damage
by gnawing the bark off trees
and shrubs.

*Bank vole leaves fine,
clear tooth marks and
unloosened bark*

*Squirrel leaves loosened
bark and some tooth marks*

Tree barked by mountain hare

Shoots stripped by bank vole

*Deer uses a tearing
action to bite
off stems*

Woodland stripped by deer

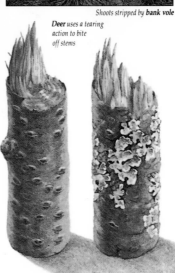

Droppings vary greatly from species to species and are one of the most significant clues to animal activity.

Hedgehog

Fox

Rat

Fallow deer

Red deer

Badger

Hare

Rabbit

Pine marten

Stoat

The squirrel is active throughout 24 hours, especially just after dawn and before sunset, and does not hibernate. The spherical nest, or drey, is about 30cm diameter and is built in the fork of a tree close to the main stem. Normally pine, spruce or larch are used. There may be several dreys in one territory. They are made mainly of grass, moss, pine cones, pine needles and bark, with buds, ferns, leaves, stems and feathers as secondary material. The inner nest is lined with soft material which is particularly thick in breeding dreys. The female looks after the young, and produces one or two litters of 1—6 young in spring and summer.

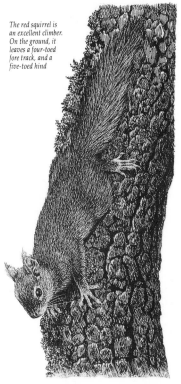

The red squirrel is an excellent climber. On the ground, it leaves a four-toed fore track, and a five-toed hind

It can be a noisy animal, making a 'chucking' call throughout the year, associated with tail flicking and foot stamping if alarmed. Food consists of a range of vegetable material. Tree seeds, beech masts, tree foliage and fungi are major items of food. Scots pine is the main food tree. The red squirrel also uses other conifers, taking nuts, foliage and shoots from broadleaved trees.

Scratch marks may make bark rough and chipped. The dreys are often more than 8m above the ground. Feeding remains are similar to those of grey squirrel. Shoots and buds are neatly 'cut', toothmarks on fungi are about 8mm across. Bark is stripped on conifer trunks, often in long spiral twists, which may have a social significance. Pine cones have the scales torn off. Droppings are small (about 6mm diameter), dark and rounded. They are widely scattered on forest and plantation floors. Specific urinating points may be used. The hind tracks have five toes (45 x 35mm) and the fore have four toes (35 x 25mm). The large

fore   hind

palm pad has three lobes on the forefoot, and four lobes on the hind. Toe and claw marks are long and fine, and hand outlines are common. Heelmarks sometimes show in tracks. The trail is normally a series of hops and bounds 45cm—1m apart.

STOAT

Mustela erminea

The stoat is a long, small member of the weasel family, distinguished by reddish-brown upper parts, white underside and tail with a black tip. The body is up to 30cm, and the tail up to 12cm. There is some size overlap with the weasel. In northern Scotland the coat turns entirely white in winter, except for the black tail tip. Further south it remains brown, or turns partially white.

The stoat kills its prey with a bite to the back of the neck

It is widespread in many habitats where there is suitable cover. The stoat will often forage by watersides and at the coast. It is found over the whole of mainland Britain and Ireland; also on many islands, being absent only from some of the Western Isles and Orkneys.

The stoat is generally solitary, although families may stay together well into the year (reports of stoats hunting in packs may reflect this). It is mainly active at night throughout the year. The inquisitive stoat, after making an initial dive for cover when disturbed, may come back to look at the quiet walker.

It can swim and climb well and spends much time exploring burrows, ditches, old walls, hedges and woodland. This species tends to be territorial, adopting a shrill bark or hiss as a threatening call.

The breeding nest may be in a litter-filled ditch, niche in a stone wall, under a hedge or in rabbit warrens. The young are born in April or May and the litter is between 6 and 12.

The stoat hunts by scent. The traditional food is rabbit, now less available since myxomatosis. It eats many birds, small mammals (especially mice and voles), rabbits and hares,

though will take carrion, earthworms and other invertebrates when food is short and will also eat eggs.

There are various signs. Carcasses with bite marks on the back of the skull (less than 15mm across) are from stoat or weasel. Feathers stripped from birds are roughly bitten through. Eggs are broken with teeth marks 8—10mm apart. Droppings are dark, elongated and irregular in shape, 40—80mm long, coiled and twisted, often with twists of fur at the ends. They have a strong musty odour when fresh, but weather quickly. They contain fur, feather and bone fragments and may be piled in or outside the den, or scattered at random. Droppings impregnated with scent are used to mark territory. Both tracks have five toes (up to 60 x 30mm in males and 45 x 25mm in females). Large toe pads have short claws and the palm pads are lobed.

BANK VOLE

Clethrionomys glareolus

In common with all voles the bank vole has rounded ears and muzzle. This small mammal has a rounded snout, russet-coloured back and a tail about half its body length (body up to 110mm, tail up to 65mm). The ears are more prominent than the field vole's.

It is found in deciduous woodland, scrub, hedgerows, banks and disturbed areas with good ground cover (eg old quarries). Smaller numbers are found in conifer plantations. It is common on road verges and ditches with hedgerows as well as dense scrub areas. The bank vole is found in larger gardens and has been recorded on open moorland where there is good ground cover from heather. It is found all over mainland Britain and some islands, but is absent from the Isle of Man, some Western Isles, the Orkneys and Shetlands. In Ireland, parts of the south-west have now been colonised.

This species is active by day and night and throughout the year. It follows regular surface runs and underground tunnels. It is frequently seen in the day by the patient naturalist who is prepared to sit quietly and wait.

The nest, which has a distinct entrance, is made of leaves, moss and feathers in woodland, and grass and moss in grassland. Breeding nests may be in tunnels or above ground in tree trunks, small rock piles and even in coarse, dense bush litter. The nest is often 2—10cm below the surface and is at the centre of a tunnel complex. Several litters of 3—5 young are born between spring and autumn.

It feeds on a wide range of buds, leaves, seeds and fruit and will also eat insects and other invertebrates, including snails. Hazel nuts are opened with neat holes, showing small teeth marks on the rim, but none on the outside (a sign of wood mice). Voles climb into canopies of willow, birch bushes and young conifer trees.

Voles are very fond of rose hips, eating the flesh, but leaving the seeds (in contrast to the wood mouse). They

also gnaw fungi, leaving small, neat teeth marks. Snails taken by a vole are nibbled along the spiral of the shell, and these may be found on quiet woodland paths.

Berries and nuts may be stored in old nests, tunnels or under stones. Tunnel systems and runs with obvious entrances (about 4cm diameter) are found in vegetation or on banks. Droppings are round in section, up to 4mm diameter and 8mm long. They are smaller than those of wood mice and lack the green colouration of field vole droppings. They are dark when fresh and left at random or in latrines. The small tracks (five toes on hind 15 x 17mm, four toes on fore 11 x 13mm) have a short heel on the hind track, which distinguishes them from mice tracks. There are short, attached claws and hand outlines are common. Tracks are unregistered and about 65mm apart.

FIELD VOLE

Microtus agrestis

The coat is rather long and shaggy, greyish-brown in colour. The ears are very small. The coat is distinctly two-coloured, paler underneath, and the tail is very short (body up to 130mm, tail up to 45mm). Rough grassland is the preferred habitat, including young forestry plantations. This vole may also be found in smaller numbers in woodlands, hedges, upland areas and coastal habitats. It occurs all over mainland Britain, but is absent from Ireland, the Isle of Man, some of the Western Isles, Orkneys and Shetlands.

It is active throughout the day and night, all year round. This species makes extensive tunnel systems in coarse vegetation and litter. There may

be many animals in a small area, but each tunnel system is used by only one vole. The nest is made of finely shredded grass, which may be in a tunnel below ground or in a grass tussock. Several litters of 4—6 young are produced from spring to autumn.

It feeds mainly on green leaves and stems of grasses, but in hard conditions will take bark. When animals meet there is a loud chattering, which is easily heard in a field in which large numbers of voles occupy the litter layer.

The main signs are the tunnel systems, which may be made in vegetation or partly between vegetation and ice or snow. Chopped grass, tunnels and piles of droppings (green when fresh), possibly marking territorial boundaries, are common signs. The tracks are similar to those of the bank vole, but slightly larger (hind 15 x 18mm, fore 12 x 15mm).

WATER VOLE
Arvicola terrestris

This is a large, long-tailed vole the size of a rat (body up to 200mm, tail up to 110mm), but distinguished by its shorter ears and muzzle. The rather coarse coat is dark, sometimes black.

It is normally found close to fresh water, provided food and cover are available. This vole is widespread over the whole of England, Wales and much of Scotland, although populations are patchy. It is absent from Ireland, probably the north-west of Scotland and all islands except the Isles of Anglesey and Wight.

Greatest activity takes place around dawn and dusk but you will often be rewarded with a sighting in the day if you sit quietly on a river bank. The vole has poor eyesight and can often be closely approached. It is active throughout the year. A great deal of time is spent in the water, but also on the bank where runway systems in bankside vegetation and laterally flattened tunnels with entrances above and below water are found. Territories are often marked by latrines. The burrow systems may

extend more than 10m from the water's edge, and contain the nests of chopped reeds, water flag and grasses. Occasionally, nests may be made on the surface in dense reed beds. Litters of 4—6 young are born throughout the summer. The water vole's food consists mainly of grasses, sedges and rushes.

Signs include the burrows, runways and prominent boulders or patches of mud where piles of chopped vegetation and droppings may be found. Areas of vegetation around the burrow entrance may be closely cropped into a 'vole garden'. The droppings are cylindrical with rounded ends, about 10—12mm long, 5mm diameter, smooth and light green-khaki in colour. Tracks display five toes on the hind (30 x 30mm), and four toes on the fore (18 x 23mm). They are distinguished from rat tracks by the short heel on the hind foot and distinct star shape to the fore. This species normally moves with a fast walk, which leaves straddled, partially registered tracks about 10cm apart.

WEASEL

Mustela nivalis

This is the smallest carnivore and member of the weasel family found in the British Isles. It is generally smaller than the stoat (body up to 23cm, tail up to 6cm). The tail is short and does not have a black tip. The back is brown and the flanks are white. The male is larger than the female. Widespread in

April to May. Four to six are born, and there may be a second litter. An agile climber, and capable of swimming well, it is active throughout 24 hours. The weasel regularly stands upright to look around and moves very quickly. Sexes remain separate with their own territories. Families of mothers and young may stay together and hunt in packs, as with stoats.

It feeds mainly on voles and mice, but also kills rats and rabbits and will take birds, nestlings and eggs from nests. Activity is indicated by feeding remains, such as bird legs, tails, and droppings outside dens. Eggs have

every type of habitat, it is often found high in the hills and in large parks in cities. Densities vary greatly from high in woodland to low in open exposed areas, and are controlled by the food supply. It is widespread over mainland Britain, but absent from Ireland, many of the Western Isles, Orkneys, Shetlands and the Isle of Man.

It rests in burrows taken over from other species or in crevices in stone walls, and does not have permanent dens, except for breeding. Breeding nests are lined with fur from the weasel itself, or its prey. Feeding remains and droppings are found outside the dens. Young are born from

teeth marks less than 6mm apart, while plucked feathers have the stems torn through. Droppings are very musty when fresh and similar in shape to those of a stoat, but smaller and thinner. They are very twisted and curled, 3—6cm long and dark, and contain fur from mice, voles and the weasel's own body. The weasel displays five toes on all tracks (hind 15 x 13mm, fore 13 x 11mm). Palm pads are delicate, claws are short and continuous with the toe pads. It moves with an arched-back gallop.

WILD CAT
Felis sylvestris

The wild cat is like a large domestic cat (body up to 65cm), but with a thick bushy tail with a blunt, black tip, distinct rings and striped body markings. It is found in open hill areas

and high woodland. Until recently habitat loss and persecution meant that the wild cat was confined to Scotland, north of the Central Lowlands. It appears to be spreading south, however, into the Southern Uplands and northern England. In some areas interbreeding with wild-living domestic cats is weakening the truly wild strains.

Mainly solitary and active at night, it is an agile climber, hunting mostly on the ground by stalking and pouncing. It follows a regular, but very large territory. The wild cat often basks in the sun and can swim when necessary. Dens are made under rocks, under tree stumps, fallen trees and occasionally in large birds' nests. Breeding dens are normally very isolated. Most kittens are born in May, but there may be a second litter in August. The food consists of small mammals, rabbits, hares, birds, frogs and sometimes fish.

There are various signs. Large prey may have the head torn off and only the brains eaten. Smaller prey may be stored, but is normally eaten when killed. Larger prey remains can be identified by chewed and partly broken bones. The droppings, musty smelling when fresh, are not buried but are deposited in regular latrines, or at random — sometimes on prominent ledges. The black/green 'scats' are twisted and may contain bone fragments, fur, feather and, sometimes, insect remains. These droppings are 4—8cm long, 1.5cm diameter. Tracks, which measure about 60 x 50mm, have large three-lobed pads with four (rarely five or six) toe prints with no claw marks. In a walking trail the tracks are in a straight line and unregistered.

DOMESTIC CAT
Felis cattus

This familiar species, which is included here out of alphabetical sequence so that it may be compared with the wild cat, is widespread everywhere and lives semi-wild in many parts of the country. Cats will kill small mammals and birds of all sizes and are now the commonest country predators.

Domestic cats are smaller than wild cats. The body markings on the tabby varieties, which are rather similar to the wild cat, are less distinct. It is important to remember that wild and domestic cats will interbreed as species.

Partially buried droppings in scrapes are a sign of the domestic cat, whereas the wild cat leaves them on the surface. Droppings are generally poorly structured and strong smelling. The tracks of domestic cats are similar to those of wild cats, but are smaller.

AMPHIBIANS

Amphibians are 'cold blooded' vertebrates with no scales on their naked skins. They breathe through gills in their larval stage but develop lungs as adults after metamorphosis. Most lay eggs in water which are then left to develop on their own. The adults live on animal food, but some larvae eat plant material. The amphibians of the British Isles fall into two groups. In one are frogs and toads, in the other are the newts.

FROGS AND TOADS

These animals have short bodies and long powerful hind legs which are used for jumping and swimming. The young, known as tadpoles, live in the water, but the adults spend most of their lives on land in damp situations to keep their skins moist. In winter they have to find cover and hibernate. The young hatch from the eggs, and then go through several stages during which they develop a tail, legs and then lose the tail to take on adult form. This process can take a few weeks or many months.

Frogs tend to spend the day sheltering in surface vegetation, whereas toads hide under wood, in crevices or burrows in the soil.

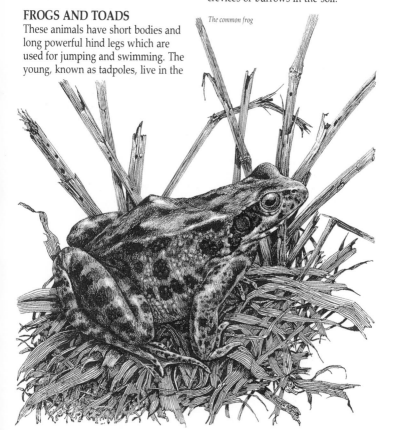
The common frog

Frog and Toad Tracks

Toads and frogs move across soft ground leaving distinctive tracks as the example of the common toad shows. The hind feet carry webbing and there are small raised areas on the toes and palms, known as tubercles, which vary from species to species, as they do in frogs. The webs of the common toad

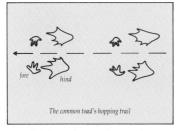

fore *hind*

The common toad's hopping trail

extend right to the toe tips, on the natterjack they go only half-way up the toes, while on the common frog they reach the second joint on the longest toe.

In the common toad the webbed hind foot shows five widely spaced toes. Only the toe impression shows sometimes, but normally at least one of the tubercles on the inner side of the palm shows well. In adult female toads the hind track may be 28 x 22mm. The four-toed forefoot is up to 15 x 15mm. In the walking trail the straddled hind feet point forward, while the forefeet are at right angles pointing in towards the median line. In the walking trail the alternately placed tracks are about 3.5cm apart. At the hop the tracks are in fours with the two hind tracks immediately behind the two fore. Small hops cover about 10cm, while long jumps cover more than 40cm. The tracks never register and there is great variation according to the underlying ground conditions.

COMMON FROG

Rana temporaria

This species, up to 7.5cm long, has smooth skin with small flattish warts. The female is a little larger than the male. During the breeding season the male develops dark callouses on the first toe of each front foot and also a bluish back. The background colour varies from yellow to black. Males tend to be darker than females. There is normally a stripe running along the inside of each foreleg. This frog is very widespread and is especially common in Ireland; elsewhere it is less common than it used to be. It is found in fields, meadows, woodlands and gardens where there are damp conditions and plenty of shade.

The common frog is active throughout most of the day, but hibernates from mid-October to late February. It usually hibernates in mud at the bottom of lakes or ponds, or in moist land. Breeding takes place in a variety of waters, from small peaty pools to large lakes. Egg laying takes place from March to April. Pairs of mating frogs, with the male on the female's back, clasping firmly with his

front feet, are a common sight at this time. The female lays her eggs in the water in large masses which first sink to the bottom and then rise to the surface. The female may produce 1,000 eggs in an hour and a total of 4,000. In favoured breeding places there will be hundreds of thousands of eggs forming a huge mass of jelly.

Frogs will migrate in large numbers to favoured breeding 'ponds'. It is now thought that adults are attracted by glycollic acid which is produced by certain algae. The development of the young normally takes about three months. Enormous numbers of young frogs make their way across the land from the breeding area. Vast numbers are killed, but those that survive return to the same place to breed when they are mature at 2—4 years old. Common frogs are normally silent animals, but make a purring croak in the breeding season, usually under water.

Adults feed entirely on land, taking insects, worms and other invertebrates. Migrating adult and young frogs are frequent road casualties. Those that survive are common prey for brown rat, water vole, otter, badger, polecat, hedgehog, water shrew, heron, hawks, owls, crows, gulls and ducks.

MARSH FROG
Rana ridibunda
This very large frog (up to 15cm) is olive green, with the head and back sometimes light green. There are largish warty bumps on the back, sometimes with black spots. There are dark green bars on the legs, and the back of the thighs are marbled with olive or whitish-grey. The underside is generally light. The iris of the eye is golden yellow.

This frog was introduced from mainland Europe in 1935 and is now widely established in Romney Marsh and Rother Levels in Kent, extending to Pevensey Levels near Bexhill and into the Isle of Sheppey in north Kent. It has been successful here because of the habitat and food supply. Romney Marsh has traditionally been a sheep farming area with small fields, broken by many ditches with strips of ungrazed grass and few hedges.

The marsh frog emerges from hibernation early in April and breeds between April and June. This is a prolific breeder and the male frogs are aggressive, with a strong migratory instinct. The male has greyish pea-sized vocal sacs in the corner of the mouth which are inflated with water to give a loud 'laughing' call.

This is a voracious frog which eats insects, worms, snails and the tadpoles of common frogs, as well as fish, lizards, snakes and mice. It enjoys basking in the sun on lily pads.

COMMON TOAD

Bufo bufo

This is a large amphibian, often up to 12cm long, with powerful front legs with which it tends to crawl rather than hop or jump. The skin is dry, dull and warty and often thick on the back, and the limbs are horny. Colour varies greatly, according to age, season, place and sex. The upperside ranges from olive to brownish-black. Older males tend to be lighter in colour.

The common toad is widespread in England, Wales and Scotland (it is the only amphibian on the Orkneys), but is absent from Ireland. It is able to tolerate dry conditions, but is not generally found at higher altitudes. During the day this toad will shelter under stones, tree roots or in dense vegetation in woods, gardens and fields. At night, or in damp weather, it crawls or hops searching for food.

After emerging from hibernation the toad spends most of its time in the water in the breeding season. The males clasp on to the female's back. The female lays as many as 7,000 eggs. These are laid in long strings of 2—4 rows (not like the large masses laid by frogs). The very small tadpoles hatch after 8—10 days and the young toads, only about 1cm long, take to the land in June or July. They become mature at four years and may live for many years. Males croak more or less continuously in the breeding system.

Toads hibernate in the same sort of places chosen for resting during the day. Toads can go for long periods without food in dry times, but cannot stand being in totally dry conditions.

Prey is carefully stalked and caught on the well-aimed sticky tongue. The food consists mainly of worms, slugs, woodlice and insects, although young snakes, lizards and frogs may also be taken. Toads are beneficial since they destroy large numbers of snails and damaging insects in gardens and crops. If they are threatened toads adopt a defensive position, inflating the lungs to make themselves look bigger, extending the hind legs and raising the back.

NATTERJACK TOAD
Bufo calamita

This is a small toad up to 8cm long, with very short back legs, which enable it to run quite quickly. Both sexes are the same size, but the male has a large vocal sac. The back is olive-green to brown, with irregular grey or red-brown blotches. The underside is pale, while a distinctive narrow, pale-yellow stripe runs along the back.

This is now a rare species restricted to a few areas. It prefers sandy places, such as coastal dune systems and some heaths. Although it can tolerate drought conditions, it prefers pools and wet patches in such areas. It has largely disappeared from much of its range in Surrey, Hampshire and Dorset and is mostly confined to north-east England. By day it shelters under stones, in small rodent tunnels, or in burrows it has dug for itself.

The natterjack comes out of hibernation in April and then moves into pools, often of brackish water, where breeding takes place in May. The eggs are laid in relatively small numbers in 1 or 2 rows twisted around water plants. They are laid at night and the tadpoles become toads after 6 or 7 weeks. The young toads are very small.

Males have a very loud trilling croak in the breeding season — indeed, this is our noisiest amphibian.

The food consists of insects and grubs, caught at night. This toad runs for short distances and rests, but can also hop. When alarmed it has the capacity to change its colour.

NEWTS

The newt has a long slender body, short head and a long tail, with a flap of skin along its lower and upper edges. The skin is very thin. The mucus-covered eggs are laid singly, or in small groups, and attached to underwater plants. The front limbs have four toes, the hind have five. Most species have tiny backward pointing teeth.

Newt Signs

Newt signs are limited by the small size of this species. Skins are frequently shed, or 'sloughed', particularly in spring, and may be left hanging among the water plants, although they are often eaten by the newt. Outside the breeding season newts spend much of their time on land and they occasionally move through an area of mud or slime soft enough to record their tracks. The tiny tracks normally show as toe marks only. In the crested newt, the largest species, the hind feet (5 x 6mm) have five toes and the forefeet (4 x 5mm) have four. Not all the toes will show and details of the rest of the foot are rare. As the animal moves across the ground the tracks point straight forward, but are widely straddled and unregistered. There is a side to side body-and-tail drag which may obscure some of the tracks.

SMOOTH OR COMMON NEWT
Triturus vulgaris

Its body is normally up to 9cm. The male is larger than the female, and has a slender body with a marked crest, while the female is heavier with a less obvious crest. The background body colour varies. The female is generally a dark-brown colour, while the male has an olive-brown back, orange belly and whitish-yellow sides outside the breeding season. In the breeding season the colours are intensified and the crest becomes very prominent in the male. The body and tail have rows of black spots, the belly becomes bright orange, the crest has dark tips and the back toes grow fringes of skin. Outside the breeding season the skin is slightly warty, but it becomes smooth when the animal takes to the water to breed.

This is the most widespread of our three newts, being found all over mainland Britain and Ireland, but is absent from the Orkneys and Shetland Isles. It is found in a wide range of

habitats. Outside the breeding season all of its time is spent on land. It feeds at night, hiding for much of the daylight hours.

In the breeding season, still water is sought out, especially weed-filled ponds. After hibernation, which takes place between about October and March, large numbers of newts can be seen in ponds, pools, lakes and ditches in certain areas. If undisturbed they

The newt is land-based outside the breeding season

will spend much time basking on the surface, but if alarmed will swim down to the bottom. The female normally lays between 100 and 200 large (3mm diameter) pale-yellow eggs. These are attached singly to water plants and may have a leaf curled around them for protection. The eggs hatch and the larvae become very large (up to 6cm long) before metamorphosing into adults after 3—4 months. Some larvae take much longer to change. The animals reach maturity at 2—4 years.

The larvae are very active and feed on water fleas, other small crustaceans (such as freshwater shrimps) and fly grubs. The adults take the same food, as well as snails, worms, tadpoles, insects and may attack other newts of the same species.

CRESTED OR WARTY NEWT
Triturus cristatus

This is the largest of our newts, up to 16cm long. The female is bigger than the male. This newt is a dark-coloured animal with a brown or black back with white dots on the sides, and an orange or yellow underside with a few dark spots. In the breeding season the male has a tall, serrated crest on the back and there is a shiny band on either side of the tail. Outside the breeding season the skin is warty,

more time in the deeper water than other newts, and in the breeding season it prefers deep-water areas with dense vegetation.

some of the skin glands producing an irritating and possibly poisonous secretion to keep some predators away. If disturbed this species may produce a very strong smell.

It is found all over mainland Britain although is absent from Ireland and many of the Scottish islands. Restricted to a limited number of ponds, this species is much rarer than it once was. It is generally found in woodland at lower altitudes. It spends

Some animals hibernate in mud under water, but the majority pass the winter under shelter on land. The 200—300 eggs are laid in April to May. Each egg is attached to the underside of a water plant leaf, which is curled around it. The larvae are up to 10mm long and feed on insects and worms. The animals become mature at two years. The adults feed on worms, snails and grubs which are usually shaken violently before being swallowed. In their turn the newts are preyed upon by larger animals.

PALMATE NEWT
Triturus helveticus

The female is slightly larger than the male and may reach 9cm. The male has a thread-like filament at the end of his tail and swollen webs on the hind feet in the breeding season, which may show in the tracks.

There is a prominent ridge along each side of the back. The dorsal crest is lower than in the smooth newt and is straight along its edge. The upper side of the male is olive-green or brown with dark-green markings and there is a stripe on each side of the head. The female is similar in colour to the smooth newt. Both sexes have a light-coloured underside though the colours tend to be darker when the animal is on land.

The palmate newt is found, with a patchy distribution, all over England and Wales. It occurs in a range of habitats from the coast to high on the hills. It prefers shallow water in the spring breeding season, but is not so specific about the type of water, and can be found in ponds, lakes, slow-flowing streams and rivers and even brackish water. It is particularly fond of a gently-sloping muddy bottom with a dense cover of vegetation.

After emerging from hibernation in March to April the female lays 300—400 single eggs, which are again attached to underwater plants. The newts are small when they metamorphose by July or August (about 25mm). This species feeds mainly on small worms, insects and their larvae.

REPTILES

Reptiles are 'cold blooded' vertebrates which have horny scales on their skin. They breathe with lungs and some lay eggs on dry land, although some species apparently give birth to live young as the eggs hatch during, or shortly after, laying. All reptiles are warmth-loving and are commoner in the south, seeking sheltered sunny positions. They will avoid direct contact with intense sun, hiding in burrows or other shelter. They have to hibernate in winter, the most southerly species hibernating the longest.

The reptiles are represented by two groups in the British Isles, the lizards and the snakes.

LIZARDS

These have five-toed limbs and an elongated body with a tail, with the exception of the slow worm which has no external limbs. Lizards have ear holes and eyelids, unlike the snakes. They move across the ground in a curving course, propelled by limbs which only move horizontally. Most are carnivores, many do not need to drink water and several species have the ability to break off their tail if attacked (the tail will subsequently grow again).

VIVIPAROUS OR COMMON LIZARD
Lacerta vivipara

The body and tail are up to 16cm long. In the male the tail is longer than the body, in the female (which tends to be a heavier animal) it is about the same length. The upper side varies from grey-brown to red-brown, broken by pale or dark spots.

Along the sides are rows of white, yellow and sometimes black spots. In the males the underside is reddish-yellow, in females it is bluish-grey. Sometimes, completely black or very dark forms occur.

The lizard is widespread over much of the British Isles, excluding the Orkneys, Shetlands and the Outer Hebrides. It is found mainly in damp places on meadows, moorland and woodland edges. *(contd pg 105)*

The common lizard is a good climber

The sloughed skin of a viviparous or common lizard

The common lizard enjoys basking in the sun

A lizard trail left in the sand

The lizard that is invariably confused with the snake. The slow worm is, in fact, a legless lizard

Tadpoles in early and later stages of development

The smooth or common newt takes to the water in the breeding season, during which time its colours intensify

The female (spotted) and male palmate newt

The natterjack toad has a very loud croak, heard in the breeding season

Stringy toad spawn is quite unlike frog spawn

Common toad

Frog spawn appears in large masses

Frogs mate during March and April

The grass snake, Britain's largest reptile, is at home on land and in water

The extremely rare smooth snake lives in dry, sandy habitats in southern England

Britain's only poisonous snake, the adder, unmistakable by its zig-zag markings

It tends to occupy a limited area, where it often basks on prominent features in the sun. If disturbed, and at night or in cold weather, it will retreat below bark, moss or other vegetation. This species cannot run very quickly, though it is an agile climber and can swim extremely well. Hibernation lasts from October to February/March. Mating takes place in May and the young, normally 6—8, but as many as 12, are born in July. The name 'viviparous' means giving birth to live young: in fact the young emerge from eggs at the time they are laid. The baby lizards are about 38mm long and once they are born the female has nothing more to do with them.

This lizard feeds mainly on earthworms, insects and grubs. It is known to eat the young of its own species. Large prey is first banged on the ground and then swallowed longways. Prey is always eaten whole. Common lizards are preyed on by foxes, martens, polecats, crows, hawks and cats.

As with other reptiles the skin is sloughed frequently, especially by the male, and is fairly commonly found.

Male and female (pregnant) common lizard

Lizard Trails

The trails left by lizards with legs are often seen in sandy areas, although the animals are very light. The trail appears as a broad body drag, with what looks like miniature oar marks on either side. Details will not show in sand, but occasionally lizards will cross muddy areas and the features become more distinct. In sand, the width of the whole trail will be between 2 and 7cm.

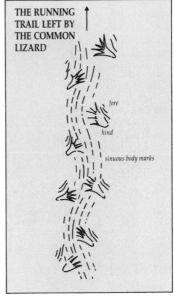

THE RUNNING TRAIL LEFT BY THE COMMON LIZARD

fore

hind

sinuous body marks

In a clear situation, such as the example of the common lizard shown here, the tracks are five-toed (about 11 x 6mm), but often only show four toes. The tracks are straddled, turn inwards, but are parallel to the median line. Tail drag is present, but almost straight and there is a suggestion of body scuff marks. There is no track registration and the stride in this running trail is between 6 and 7cm.

SAND LIZARD
Lacerta agalis

The body and tail are up to 20cm long. The male has stouter hind legs and a thicker tail base than the female. Males are a bright green, females are browner and both have a more bulky head than the common lizard. The claws on the forelimbs are much longer than those on the hind.

This very rare reptile has a very restricted distribution in southern England. It lives on open heathland, stone walls and sand dunes. The sand lizard will dig its own burrows, but often takes over old mole, vole or shrew runs, the entrances to which are blocked with vegetation or soil as a protection against cold. Hibernation takes place from September to April, the males emerging before females.

There are violent, biting fights between males over females before mating. Mating takes place in May and the eggs are laid about six weeks later. The female becomes very large before the eggs are laid. The 5—13 eggs are laid in a burrow 6—7cm deep, dug by the female. The eggs hatch after 6—8 weeks, and the young have to fend for themselves immediately.

This lizard feeds on insects and other invertebrates in the tangled stems of older heather bushes and will also eat slugs, snails and worms as well as carrion.

SLOW WORM

Anguis fragilis

This legless lizard is up to 50cm long. The body is tubular and there is no marked break at either the head or the tail junction. There is no external sign of the ear cavities. Male slow worms are bronze all over, but the females and young have a dark belly, which contrasts with the pale-brown back.

There is a black dorsal stripe which becomes less prominent with age. Blue-spotted males sometimes occur. On rough ground the animals move with side to side undulations, while on smooth ground they will pull themselves forward with the chin and backward with the tail.

The slow worm is found all over mainland Britain, but is more common in the south, and is absent from many Scottish islands. It prefers well-vegetated sites which are warm and have good ground cover. These include deciduous woodlands, rough pasture, heathlands, hedgerows, marshland and gardens. The slow worm is particularly fond of damp places.

This is generally a secretive species which comes out mainly at dusk to search for earthworms, one of its main sources of food. It lives under stones, fallen wood, corrugated iron sheets or in burrows. These are also used for hibernation and it is not unusual for 30—50 animals to share the same burrow, which may be blocked with soil for protection against cold. Hibernation takes place from October to early April and mating occurs soon after emergence. The young are born in August or September (it produces 5—25 young, but 10 are most usual). The young are up to 9cm at birth, hatching from eggs at the time of laying.

The slow worm searches for prey with deliberate movements, from dusk onwards or after rain, looking for worms, slugs, snails, insects and grubs which live in the soil, litter and under logs or stones. There may sometimes be a tug of war between two slow worms which have hold of the same large worm. In their turn these lizards are preyed on by hedgehogs, other mammal carnivores and birds of prey.

The skin is sloughed off in several sections every six weeks or so and, in common with the other lizards, slow worms shed their tails in an emergency (but the tail does not regrow).

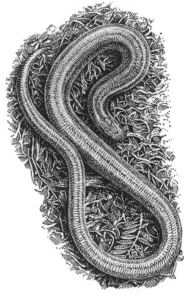

SNAKES

Snakes have an elongated body with a relatively short tail and no limbs. The rib cage is not joined to a breast bone and the end of the ribs and a series of muscular contractions are used to move the body across the ground. Normally, as the body is pushed forward on a curving course, it leaves a faint sinuous furrow (if the ground, such as sand, is soft enough to record it), which will vary in size with the species and age of snake.

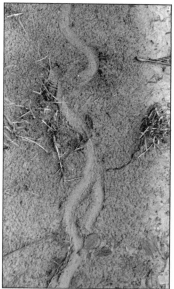

Snakes usually mark their passage over soft surfaces

The jaws and face bones are loosely hinged so that relatively large prey can be swallowed. There are slender backward-pointing teeth on the upper jaws and the adder has fangs to inject venom. Sloughing of skin takes place almost monthly, from shortly after hatching until hibernation.

ADDER

Vipera berus

This is Britain's only poisonous snake, normally between 50 and 60cm long, but occasionally up to 100cm. The female is often larger than the male. The body is broad and the head is flat.

The eye is reddish and the tongue, which slips in and out of the small gap between the lips, is black. The adder's background colour varies, but the males tend to be pale grey with black markings, while the females and young snakes are browner with less obvious markings. At the back of the head is a dark 'V' or cross-shaped marking, but the distinctive feature of this species is the dark zig-zag line which runs all the way down the back.

The adder is widespread over mainland England, Scotland and Wales but is absent from Ireland and many of the Scottish islands. It is found mainly in dry habitats, such as heathland, moorland, sand dunes, the edge of open woodland, stone walls and old quarries, as long as there are places to bask in the sun.

Adders hibernate from October to March or April, making use of holes in the ground, tree roots and crevices in walls. Snakes living on open moors and heaths may have to travel considerable distances to find suitable habitats and often large numbers of animals will use the same place. Mating occurs in May or June, and there are violent battles, 'the dance of the adders', between males over females at this time. The young, born in egg sacks which break immediately, appear in August or September and number 5—15. They take 4—5 years to mature as they grow slowly. In summer a great deal of time is spent basking, and resting adders are a common sight on some heaths. This is one of the few reptiles that is so obviously about throughout the 24 hours of the day. Shelter consists of crevices, stones and roots although mouse burrows may be pressed into service after the occupants have been eaten. Although they are competent swimmers they generally avoid very wet areas.

Adders feed on lizards, frogs, mice and the eggs and young of small birds. Prey is killed by a strike from the poison fangs, after it has been located by the tongue which picks up scent, temperature and vibration. As many as 30 mice may be eaten in a summer, but digestion is a slow process. Adder bites can be unpleasant and occasionally fatal to human beings, so it is important to seek medical attention if bitten, although the quantities of poison (0.1g) are intended to deal with very small prey.

Adders have many enemies, not least man. They are eaten by crows, gulls, birds or prey and hedgehogs. Hedgehogs are not immune to adder venom, but they protect themselves by rolling into a ball, then rapidly uncoiling and give the adder a bite, continuing the process of biting and coiling until the snake is dead. The body is then systematically mutilated to break the bones and to make it easier to eat.

GRASS SNAKE
Natrix natrix

Measuring up to 150cm when fully grown, this is the largest British reptile. The body is long, slender and slightly compressed. The snout is flat and rounded, the eye is black. There are many different colour forms, and the upperside may be black-brown, grey, olive or reddish, but the top and front part of the head are always black.

The lower jaw is a pale white and this continues behind the back of the head, stopping just short of the spine so that, seen from above, there are two half-moon shapes.

This snake is found all over mainland England and Wales and the islands, but is absent from Ireland and the Scottish islands. It is found in grassland, open woodland, hedgerows, marshy areas, moorland and along the banks of rivers and lakes. Never far from water, it is an extremely agile swimmer, bending the body from side to side with the head held high. It hunts actively for its main prey, which is frogs. Because of this the snake is highly dependent on the maintenance of Britain's wet areas. It will feed on other wetland species, such as newts and even toads, but it cannot cope with the small mammals which are important to the other snakes.

In October grass snakes normally leave the damper areas and go into hibernation in better-drained places. They emerge in early April and, after a short time basking to warm up, mate any time from April to June. The female lays her eggs (between 10 and 20) in piles of leaves, compost heaps or rubbish tips, which give off heat and help the sticky, greyish-white oval eggs to hatch. The eggs quickly become tough and leathery after laying. Because places suitable for egg laying are limited, many females may congregate in a good area and hundreds of eggs may be laid in a single large compost heap. The eggs normally hatch in September and the black young (about 15—18cm long) can feed themselves immediately on young fish, slugs and earthworms. Grass snakes are mature at 3—4 years old. They slough their skin 5 or 6 times in a summer and the skin is often found nearly intact.

Grass snakes are vulnerable. The adults are persecuted by man, and the eggs are taken by stoats, weasels, rats and other small mammals. It is a creature which hunts chiefly by day, using sight and smell, for it is not adapted to seek out nocturnal small mammal prey.

SMOOTH SNAKE
Coronella austriaca

Adults of this slender snake, with smooth shiny scales, may grow up to 75cm long. The female is larger than the male. The head is rather small, the eyes have a yellow iris, and the tail is relatively short. The general colour of the upper surface is usually brown, but may be grey, yellowish or even reddish. The distinguishing feature is the dark 'crown' marking on the top of the head. There may be spotting and striping on the body. The underside is reddish in the young, but is generally lighter in adults and often has dark spots. This snake is now extremely rare and occurs only in parts of Dorset, Hampshire, Sussex and Surrey on the sandy heaths. It requires dry, well-drained sandy banks with mature heather bushes. It is also found in sand dunes.

The smooth snake is most active by day, favouring sunny places, although it does not go very far from its shelter, other than when hunting prey. It is a non-poisonous species, although it will bite if attacked, and feeds mainly on lizards and small snakes, occasionally taking mice and small birds. It coils around prey, which it captures by the head, but does not suffocate. The smooth snake's prey is often swallowed slowly.

Hibernation occurs from October to April and mating takes place immediately after. The young are born in August or September, rupturing the egg membrane immediately after laying. Three to fifteen young are produced and these are up to 20cm long at birth.

This species is under considerable threat because of the delicate nature of its habitat. It can swim well, but does not climb, cannot move very quickly and is inclined to range over a very small area. It has the same mammal and bird predators as the other snakes, but its real problem is the loss of breeding and feeding grounds.

FOUR-TOED MAMMAL TRACKS
(approximately ½ life size)

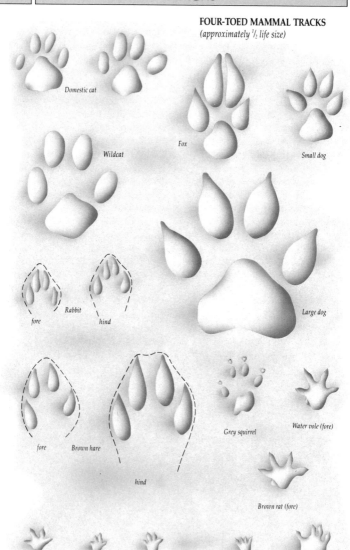

Domestic cat

Fox

Small dog

Wildcat

Rabbit
fore | hind

Large dog

Brown hare
fore | hind

Grey squirrel

Water vole (fore)

Brown rat (fore)

Wood mouse (fore) Bank vole (fore) Field vole (fore) Hazel dormouse (fore) Edible dormouse (fore)

FIVE-TOED TRACKS, SMALL- AND MEDIUM-SIZED MAMMALS
(approximately life size)

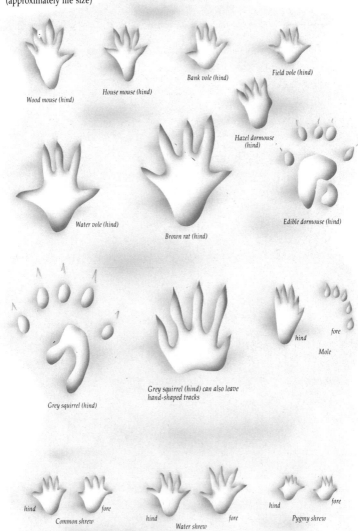

Wood mouse (hind)

House mouse (hind)

Bank vole (hind)

Field vole (hind)

Hazel dormouse (hind)

Water vole (hind)

Brown rat (hind)

Edible dormouse (hind)

Grey squirrel (hind)

Grey squirrel (hind) can also leave
hand-shaped tracks

hind   fore

Mole

hind   fore

Common shrew

hind   fore

Water shrew

hind   fore

Pygmy shrew

FIVE-TOED TRACKS, MEDIUM- AND LARGE-SIZED MAMMALS
(approximately ½ life size)

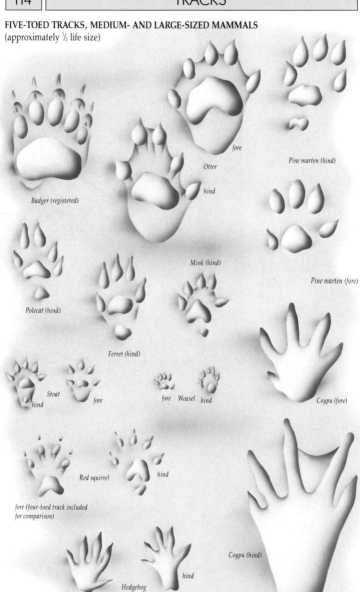

Badger (registered)

Otter — fore, hind

Pine marten (hind)

Mink (hind)

Pine marten (fore)

Polecat (hind)

Ferret (hind)

Stoat — hind, fore

Weasel — fore, hind

Coypu (fore)

Red squirrel — fore (four-toed track included for comparison), hind

Hedgehog — fore, hind

Coypu (hind)

TRACKS WITH SLOTS (approximately ½ life size)

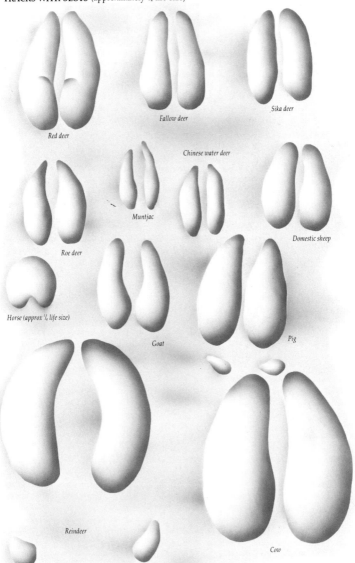

Red deer

Fallow deer

Sika deer

Chinese water deer

Muntjac

Roe deer

Domestic sheep

Horse (approx ¼ life size)

Goat

Pig

Reindeer

Cow

ANIMAL TRAILS

Mole — walking trail (approx ¹/₂ life size)

Pipistrelle bat — walking trail (approx ²/₃ life size)

Common seal — flopping at speed

Crested newt — running trail (approx life size)

Common toad — hopping trail (approx ¹/₂ life size)

Common lizard — running trail (approx ¹/₂ life size)

SITES TO VISIT

Some creatures are widespread and easily seen; others have to be looked for very carefully. The list below identifies some of the many parts of Britain which are special for wildlife in some way. This may be because they are remote, or because they contain interesting species, or both. This list is by no means exhaustive: almost every part of the countryside or town has fascinating wildlife, if you know where and how to look.

The Wash, *Lincolnshire and Norfolk.* Large estuary of 20,000ha. Breeding common seals.

Epping Forest, *Essex.* Ancient woodland of beech, oak, hornbeam. Open access area. Badger, fox, fallow deer, grey squirrel.

New Forest, *Hampshire.* Heath areas, deciduous woodland and coniferous plantations. Red, fallow, sika and roe deer. Ponies. Adder, grass snake, smooth snake.

Box Hill, *Surrey.* Chalk grassland and scrub. Badger.

Box Hill

Breck Heathlands, *Norfolk.* Heather heath and grassland. Rabbits, red squirrel.

Norfolk Broads. Flooded medieval peat cutting. Coypu.

Dartmoor, *Devon.* Granite moor. Ponies, hazel dormouse, bats.

Exmoor, *Somerset.* Sandstone grasslands and heather moor. Red deer, ponies.

Tregaron Bog, *Dyfed.* Raised bog. Polecat.

Skomer Island, *Dyfed.* National Trust, restricted access. Breeding grey seal and distinct race of bank vole.

Ainsdale Dunes, *Lancashire.* Full range of dunes from foreshore to pine woods. Red squirrel, rare sand lizard and natterjack toad.

Grizedale Forest, *Lake District, Cumbria.* Coniferous plantations and moorland. Deer.

Kielder Forest, *Northumberland.* Conifer plantations and open moor. Nature trails. Deer, pine marten, squirrel.

Aviemore, *Highland Region.* Reintroduced reindeer.

Rannoch Moor, *Tayside Region.* Open moor. Wild cat, otter.

Rothiemurchas Forest, *Inverness, Highland Region.* Semi-natural pine forest. Pine marten, wild cat, red squirrel, red and roe deer.

Rhum (National Nature Reserve island), *Highland Region.* Managed for red deer.

Beinn Eighe, *Highland Region.* Western pine woods with holly and mountain ash. Red deer, roe deer, pine marten, red squirrel and wild cat.

National parks and forest parks are all worth visiting. Details can be obtained from the Countryside Commission and Countryside Commission for Scotland (see next page).

ORGANISATIONS

The following organisations can provide information on the animals and places to visit. If writing to them for information, please enclose a stamped, addressed envelope.

Countryside Commission, *John Dower House, Crescent Place, Cheltenham, Gloucester GL50 3RA.*

Countryside Commission for Scotland, *Battleby, Redgorton, Perth PH1 3EW.*

Field Studies Council, *Preston Montford, Montford Bridge, Shrewsbury SY4 1HW.*

Forestry Commission, *231 Corstophine Road, Edinburgh EH12 7AT.*

Mammal Society, *Harvest House, 62 London Road, Reading RG1 5AS.*

Nature Conservancy Council, *Northminster House, Northminster, Peterborough PE1 1UA.*

Royal Society for Nature Conservation, *The Green, Nettleham, Lincoln LN2 2NR.*

Royal Society for the Protection of Birds, *The Lodge, Sandy, Bedfordshire SG19 2DL.*

FURTHER READING

This brief list gives some key titles which will enable the reader to find more details about animals and their signs and may lead on to further reading.

Brown, R.W and Lawrence, M.J. *Mammals' Tracks and Signs. The Nature Detective Series.* Macdonald, London (1983).

Brown, R.W, Lawrence, M.J and Pope, J. *Country Life Guide to the Animals of Britain and Europe; their tracks, trails and signs.* Country Life, London (1984).

Brown, R.W, Ferguson, J, Lawrence, M.J and Lees, D. *Tracks and Signs of the Birds of Britain and Europe. An Identification Guide.* Christopher Helm, London (1987).

Corbet, G and Ovenden, D. *The Mammals of Britain and Europe.* Collins, London (1980).

Frazer, D. *Reptiles and Amphibians in Britain. The New Naturalist.* Collins (1983).

Hvass, H. *Reptiles and Amphibians.* Blandford Press (1975).

Sherwood Forest

GLOSSARY

Accessory Marks. Marks left in the ground by spines, fur, feathers, tail or body of an animal.

Aquatic. Living in or around freshwater.

Canopy. The upper branches of a tree.

Carnivore. An animal, or plant, that lives on flesh.

Cleave. The toe tip of a deer, sheep, goat, pig or cow foot.

Clocoa. Common opening of intestine, urinary tract and reproductive tract in amphibians, reptiles and birds.

Cover. Vegetation used as a shelter by an animal going about its daily activity.

Crottie. Mass of individual droppings from deer, sheep or goats. The droppings adhere together.

Deciduous. Tree or plant which sheds its leaves annually.

Delayed Implantation. The process whereby a fertilised egg does not start to develop immediately.

Dew Claws. Small, residual toes high on the limbs of ungulates.

Digitigrade. An animal which walks on the tips of its toes.

Feral. Domestic species living and breeding in the wild.

Gregarious. Animals which mix with others of their own species.

Habitat. The environment in which an animal or plant lives.

Herbivore. A plant eater.

Hibernation. A period of dormancy by which animals survive the cold and lack of food in winter.

Insectivore. An animal or plant which feeds on insects.

Latrine. A pit or raised point on which an animal regularly leaves its droppings.

Litter. New-born family of certain mammal species.

Litter Layer. Top layer of soil consisting of fallen leaves, grass and other rotting vegetation.

Median Line. An imaginary line which runs between tracks in a trail left by an animal.

Metamorphosis. The change in structure and form when a larval form changes into an adult one, eg a tadpole into a frog or toad.

Omnivore. An animal which eats both plant and animal foods.

Pedicle. Bony prominence on which the male deer's antlers develop.

Pellet. The undigested hard remains from plant or animal food regurgitated by birds of prey, gulls, crows, waders and other birds.

Plantigrade. An animal which walks on the flat of its feet, leaving toe, palm and heel pad impressions.

Prehensile. A tail which is highly mobile and can be used to help an animal grip in a tree, eg dormouse.

Registration. The process by which tracks from hind feet are partially, or totally placed in the tracks of the forefeet.

Run, Runway. Regular route followed by an animal, often a distinct path in vegetation.

Rut. The mating period of deer and other ungulates.

Slot. Half of a track made by a deer, sheep, goat, pig or cow.

Slough. Peel-off skin, as in a snake or lizard.

Species. The smallest natural sub-division of animals or plants. Individuals will interbreed.

Straddle. The distance across the median line between tracks.

Stride. The distance between two successive tracks from the same foot, eg between the left fore.

Tracks. The prints left by animal feet in soft ground.

Trail. The tracks, accessory marks and signs of disturbance left in the ground by a passing animal.

Unguligrade. An animal which walks on the tips of greatly elongated finger bones.

INDEX